W9-AYR-633

Garbage

Look for these and other books in the Lucent Overview Series:

Abortion
Acid Rain
Adoption
Advertising
Alcoholism
Animal Rights
Artificial Organs
The Beginning of Writing
The Brain
Cancer
Censorship
Child Abuse
Cities
The Collapse of the Soviet Union
Dealing with Death
Death Penalty
Democracy
Drug Abuse
Drugs and Sports
Drug Trafficking
Eating Disorders
Elections
Endangered Species
The End of Apartheid in South Africa
Energy Alternatives
Espionage
Ethnic Violence
Euthanasia
Extraterrestrial Life
Family Violence
Gangs
Garbage
Gay Rights
Genetic Engineering
The Greenhouse Effect
Gun Control
Hate Groups
Hazardous Waste
The Holocaust
Homeless Children
Illegal Immigration
Illiteracy
Immigration
Memory
Mental Illness
Money
Ocean Pollution
Oil Spills
The Olympic Games
Organ Transplants
Ozone
The Palestinian-Israeli Accord
Pesticides
Police Brutality
Population
Poverty
Prisons
Rainforests
The Rebuilding of Bosnia
Recycling
The Reunification of Germany
Schools
Smoking
Space Exploration
Special Effects in the Movies
Sports in America
Suicide
Teen Alcoholism
Teen Pregnancy
Teen Sexuality
Teen Suicide
The UFO Challenge
The United Nations
The U.S. Congress
The U.S. Presidency
Vanishing Wetlands
Vietnam
World Hunger
Zoos

Garbage

by Eleanor J. Hall

Library of Congress Cataloging-in-Publication Data

Hall, Eleanor J.
Garbage / by Eleanor J. Hall.
p. cm. — (Lucent overview series)
Includes bibliographical references and index.
Summary: Examines various methods of handling waste products, including landfills, incineration, recycling, and reducing the amount of garbage produced, as well as future possibilities for waste management.
ISBN 1-56006-188-X (alk. paper)
1. Refuse and refuse disposal—Juvenile literature. [1. Refuse and refuse disposal.] I. Title. II. Series.
TD792.H35 1997
363.72'8—dc21 96-49422
CIP
AC

P.O. Box 289011, San Diego, CA 92198-9011
Printed in the U.S.A.

Contents

Introduction

DURING THEIR LIFE cycles, all living things produce waste materials of one kind or another. Most of these natural wastes are biodegradable, meaning they can be broken down by nature and reused for other purposes.

With the rapid advance of science in recent times, this natural balance has been upset. New materials that nature cannot decompose or reuse, such as plastics and synthetics, have been created. Even many biodegradable wastes that humans produce in their everyday lives—food scraps, paper products, lawn trimmings—are discarded in such quantities today that natural decomposition processes cannot recycle fast enough to manage the mounting piles of refuse.

The result is that cities and towns all over America are faced with ever-increasing amounts of waste products from homes, offices, restaurants, stores, schools, and hospitals. The Environmental Protection Agency (EPA) calls these materials "municipal solid waste." More commonly referred to as garbage, these "everyday" wastes have become a major headache for local governments. As reported in the *Statistical Abstracts of the United States, 1995*, there has been a steady increase in municipal solid waste from 1960 (the first year statistics were collected) through 1993 (the latest figures available). In 1960, 87.8 million tons were generated. By 1993, that figure had risen to 206.9 million tons, a 136 percent increase in only 33 years. Since the population was also growing during that period, a certain amount of increase is to be expected. But population growth alone cannot account for the rise. In 1960 the average amount of waste generated per person per day was 2.7 pounds (calculated as total pounds of garbage divided by population divided by 365 days). In 1993 that figure had increased to 4.4 pounds per person per day. Obviously, waste is increasing faster than population.

These figures leave little doubt that a garbage problem exists in America. But the United States is not alone. The same thing is happening in other industrialized nations throughout the world. Waste generation in developing countries is rising, too, as industrialization moves forward and population increases in those nations.

Those who are faced with solving garbage problems today have only four basic methods available to deal with waste. They can bury the garbage, burn it, recycle it, or reduce the amount of garbage produced. These management techniques are not newly devised responses to the problem. They are the same methods used by the very first human beings on earth. But although the methods are

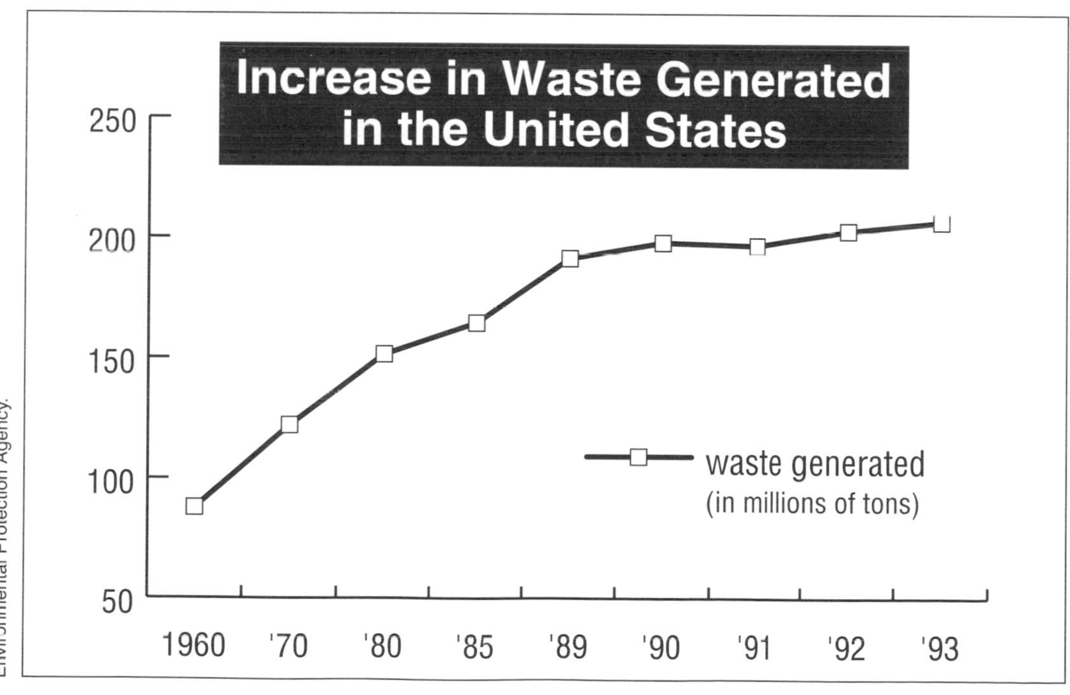

Source: Franklin Associates, Ltd., Prairie Village, KS, *Characterization of Municipal Solid Waste in the United States: 1994*. Prepared for the U.S. Environmental Protection Agency.

the same, the technology (tools and knowledge) used to practice them has changed enormously. For example, garbage is still buried, but now it is concentrated in sanitary landfills with the aid of bulldozers and dump trucks. Garbage is still burned, but in gigantic incinerators designed for that purpose.

Furthermore, technology is not perfect. Waste management methods involving burying, burning, and recycling have serious shortcomings as well as benefits. Landfills sometimes leak pollutants into the water supply, incinerators pour smoke and harmful gases into the air, and some recycled materials have no market value.

Public attitudes play an important role in the pace of progress. Many people close their eyes and refuse to admit that a problem exists. Others recognize the crisis but, understandably, resist having landfills or incinerators built in their neighborhoods. An even greater obstacle is entrenched consumer habits and attitudes: Encouraged by advertising, Americans eagerly buy the latest in consumer goods whenever they can, from hot tubs to plastic toys.

Hopeful signs

Within this gloomy picture, however, there are a few signs of hope. Recycling practices and efforts to control the amount of garbage produced have given the average citizen the means to affect waste management. "In the year 2000, Americans may be contributing measurably less to the municipal solid waste stream than they did in 1993," a 1995 article in the *EPA Journal* states. "If this happens as projected, it will be the first drop since 1960, the first year for which comparable data were analyzed."

Another favorable sign is that many promising experimental technologies to replace or supplement older ones are being developed and tested. Meanwhile, scientists and other professionals whose job it is to deal with waste disposal are continuously working to improve present-day disposal methods and make them safer.

Though vastly different, all waste management methods in use today are interrelated. Changes in one of them affects the others. A reduction in recycling programs, for example, may increase a need for landfills. Decisions regarding waste disposal, however, require consideration of the amount and type of garbage that is generated, the most efficient way of dealing with that garbage, and the effects that the disposal method may have on the community and the environment. For these reasons, towns and cities investigate a variety of options and choose waste management programs based on the needs and desires of the populace. Waste management in most communities, therefore, is not a matter of choosing one method over the others, but in finding a proper balance among all available choices.

1

Burying Garbage in Sanitary Landfills

TRAVELERS APPROACHING ST. LOUIS, Missouri, from the east on Interstate 70 may notice a flat-topped mound rising a hundred feet above the ground on the left side of the highway. The mound is an earthen monument built by the Mississippian Indians hundreds of years ago, now designated a World-Class Heritage Site.

Just a mile or two farther down the highway, travelers cannot help but notice another mound, much larger and of more recent origin. This mound is a landfill for the nearby metropolitan area, a man-made monument of garbage. While only a few places have world-class Indian mounds, every large city in America has its own garbage mound (or mounds), smaller versions of which appear even in less populated cities and towns.

As unpleasant as these landfills may be, they are an improvement over the open garbage dumps of the recent past. As late as 1965, garbage was still being dumped into holes in the ground or on open fields, usually outside city limits. In most localities, there were few limitations on what could be thrown away at a dump. The assortment included discarded items from homes, garages, and places of business such as car parts, tires, furniture, broken dishes, clothing, shoes, appliances, bottles, cans, boxes, newspapers, and magazines.

Mixed in with the junk were organic substances such as food scraps, leaves, yard trimmings, and even dead animals. Other, more potentially harmful materials—including motor oils, pesticides, shoe polish, paint, solvents, cleansers, construction debris, and old batteries which leaked acids—were also present at most dumps.

Leachate, a foul-smelling liquid derived from organic and chemical wastes, seeped out of the garbage into the ground. Flies and other insects swarmed around the rotting food scraps, rats could be seen scampering about, and fires burned here and there, giving off billows of black smoke. Occasionally, explosions occurred caused by methane, a colorless, odorless, flammable gas produced by rotting garbage. The dump was a place to be avoided, and most people did just that.

The garbage problem

Although the garbage dumps were unhealthy and unsightly, garbage was not regarded as a national problem prior to the 1960s. During the Great Depression of the 1930s many people barely had enough on which to live. Because they had very little money to buy things, people recycled and reused what they had. No one at that time considered recycling as a way of controlling waste. Efficient consumption was simply a necessity.

"SOMEDAY, SON... ALL THIS WILL BE... A LANDFILL!"

The depression had barely ended when America entered World War II in 1941. All production was geared toward the war effort. Consumer goods were rationed, meaning individuals could buy only limited quantities of certain items such as cars, tires, shoes, and gasoline. People had to make do with what they had.

After the war, life in America began to change. With neither a war nor a depression going on, people could finally buy the things they wanted. And there were so many more choices. Plastics and synthetics first developed for military purposes proved useful materials for a vast new assortment of consumer goods.

At the same time, the population was both steadily increasing and relocating to big cities. When they got there, those who could afford it built their homes in nearby suburban communities and drove cars or rode commuter trains to work.

Before long, this combination of social and economic factors created a garbage problem that could not be ignored. Not only was there more garbage being generated each year (much of which, like plastics, was not biodegradable), but, as suburban settlements spread outward from the cities, there were fewer and fewer places to put it.

Solid Waste Disposal Act

In 1965, Congress passed the Solid Waste Disposal Act, the first federal statute ever to deal with this problem. The goals of the act were to help states formulate statewide waste disposal plans and to assist local communities in replacing open garbage dumps with safer disposal areas called sanitary landfills. The act authorized grants for research and experimentation, and even provided partial funding for landfill construction.

Under regulations set forth by the Solid Waste Disposal Act, landfills were to be located in environmentally safe places where decaying garbage would not pollute streams or groundwater. Much stricter methods of treating garbage were required, as well. For example, raw garbage had to

Until the 1960s waste was deposited in open dumps with little regard for potential health hazards. The Solid Waste Disposal Act of 1965 called for sanitary landfills to be located in environmentally safe areas and capped with soil daily.

be covered daily with a layer of soil to reduce the spread of disease.

Whenever possible, local governments were allowed to upgrade existing dumps to sanitary landfill status instead of bearing the cost of building new landfills from scratch. Dumps that were located in environmentally sensitive areas or could not be brought up to the new standards were to be closed within a specified time. The general content of landfills did not change much, however, since no distinction was made in the Solid Waste Disposal Act between harmless and potentially dangerous wastes.

For several years after the passage of the new regulations, compliance lagged in many communities, sometimes because of delays or lack of interest on the part of local officials, but more often because even with federal assistance a community lacked the funds to upgrade old dumps or open new ones. A study of a garbage disposal area near Helena, Montana, in 1973 illustrates many of the unhealthy conditions that still existed throughout America eight years after the Solid Waste Disposal Act was passed. After examining a local landfill, the study commission reported:

> The present operation of the Scratch Gravel Landfill demonstrates little regard for pollution control or aesthetics [appearance]. . . . Because the refuse is not covered daily, Scratch Gravel should really be considered as a nonburning open dump and not a sanitary landfill. Due to the water table in the area it is a potential if not an actual source of pollution to the surface and groundwater. Since the refuse is not covered daily it is a source of food and shelter for vermin [rodents and insects]. The animal disposal area is an actual health hazard. . . . The Scratch Gravel Landfill cannot be considered anything less than a blight of the landscape.

In spite of a slow start in many communities across America, Congress continued to pass additional waste disposal legislation. In 1970 the Environmental Protection Agency was created. Its purpose was to coordinate all federal efforts on behalf of the environment, including waste disposal. Also in 1970, the Resource Recovery Act was passed, the first legislation promoting recycling as well as disposal of wastes.

The motor oil and paint supplies in these garbage cans are examples of hazardous waste. Early landfills that contained hazardous waste were threats to the environment and people's health.

Hazardous and toxic waste legislation

None of the early waste disposal legislation dealt specifically with problems of hazardous wastes. Hazardous wastes are substances in any form (solid, liquid, or gas) that threaten the well-being of the environment or the health and safety of living things.

The EPA categorizes hazardous wastes as ignitable (flammable waste, such as paint removers and oils), corrosive (acid-like substances, such as cleaning solutions), reactive (waste that may explode or give off deadly fumes when mixed with water, such as old weapons, ammunition, and certain chemicals), radioactive (waste materials from nuclear plants and research labs), toxic (poisonous wastes, such as lead, mercury, and arsenic), or infectious (used hypodermic needles, for example, that may spread disease).

This trash has been illegally dumped in an abandoned lot. Even toxic waste was sometimes dumped in nondesignated areas, prompting hazardous waste disposal regulation in 1976.

Some hazardous materials, such as household cleaners, pesticides, batteries, and motor oil, had been routinely discarded for years without causing public alarm. That began to change, however, when certain extremely dangerous waste products began showing up at disposal sites, often in large quantities. These included radioactive waste, agricultural pesticides, and toxic chemicals used in mining, manufacturing, and construction. An even more frightening discovery was the practice of "midnight dumping," in which toxic waste was secretly dumped or buried outside regular disposal areas by unscrupulous companies or individuals.

In response, Congress passed a far-reaching amendment to the Solid Waste Disposal Act in 1976 entitled the Resource Conservation and Recovery Act (RCRA). As its title implies, this legislation added conservation of natural resources to the goals of waste management. It also strengthened and expanded the 1965 regulations pertaining to domestic waste, and, for the first time, made a distinction between hazardous and nonhazardous wastes. Different sets of regulations for handling and disposing of each type were set forth in detail.

The hazardous waste provisions of the RCRA, however, contained no provisions for cleaning up toxic waste sites that had been created prior to the act. One such site was an old toxic waste dump covered over in the 1950s at Love Canal, New York, but not discovered until 1977 after many people became seriously ill. Responding to public outrage over the subsequent protracted Love Canal scandal, Congress passed the Comprehensive Environmental Response, Compensation, and Liability Act (CERCLA) in 1980. Better known as the Superfund, this act set aside \$1.6 billion to clean up the worst toxic sites in the country without having to wait for the courts to determine who was at fault.

A sign warns people to stay away from the filled-in dump site at Love Canal, New York. The site, an old toxic waste dump, was discovered in 1977 when several people became ill.

Once fault was established, however, polluters could be forced to repay the government for cleanups funded with Superfund monies. This feature of the law triggered powerful resistance from business and industrial organizations, as well as various political groups opposed to the act as excessive government interference. Turmoil over Superfund legislation delayed the start of cleanup work at polluted sites for many months.

In 1984, impatient with foot-dragging over hazardous waste disposal, Congress passed the Hazardous and Solid Waste Amendments to the RCRA. These amendments further clarified the distinction between hazardous and nonhazardous wastes. In addition, small businesses generating hazardous wastes (such as dry cleaners and small gas stations) were brought under RCRA regulations for the first time. The EPA was ordered by Congress to vigorously regulate and monitor the handling of hazardous wastes and to enforce the rules more aggressively.

Revised solid waste landfill regulations

For years, public attention had been focused on toxic waste dumping scandals and on political battles over toxic waste cleanups. In April 1994 attention shifted back to problems of domestic waste when the EPA published a revised set of regulations for municipal solid waste landfills.

According to the new rules, sites could not be located within earthquake zones, wetlands, floodplains, or near airports. (The reason for the exclusion of airport vicinities is that birds attracted to landfills may cause plane crashes.) Other regulations included spreading a two-foot layer of clay over the bottom of the landfill pit, double lining the pit with heavy plastic, installing a system of pipes to drain off leachate, checking groundwater for contaminates, and covering garbage daily with six inches of soil.

Because municipal solid waste landfills were not designed to handle hazardous wastes, the new rules required landfill operators to make random checks of incoming garbage for toxic materials. Installation of methane gas monitoring systems was also required in new landfills. If methane problems arose at old landfills, remedial action had to be taken immediately.

In 1991 a county in Georgia found out just how troublesome methane gas seeping from an old landfill could be. Unknown to home buyers, a housing subdivision near Savannah was built on an old, unvented landfill. "The problem became apparent . . . when a resident discovered his doors and windows would not shut," a 1992 article in *American City & County* states. "Then eaves started falling off the house, and a three-inch fissure [crack] developed in the floor." An investigation traced the problem to unsafe levels of methane seeping from the old landfill.

No one wanted to accept responsibility for the problem, but after months of arguments, accusations, and lawsuits, the county agreed to buy thirty-two of the forty-four homes in the subdivision. The affected houses were moved and a gas venting system was installed to protect those remaining. Even after the displaced houses were resold, the county suffered a huge financial loss.

Hazardous waste landfill regulations set in 1994 included limiting the locations of the sites and lining the pit with heavy plastic to insure against leakage and contamination.

When landfills are closed, site operators must cap them with eighteen inches of water-resistant soil and top that with six additional inches of earth for replanting. Closing rules even extend into the future. In the August 1993 issue of *Governing*, environmental writer Tom Arrandale reports, "Before a landfill opens, operators must have plans for closing it down when all of its space has been filled. Owners must submit plans for monitoring gas, collecting leachate, and preventing the cover from being disturbed for 30 years after the last waste is dumped."

Well before the latest set of rules, many large cities had begun to upgrade their landfills. Nevertheless, even for them, and certainly for many smaller cities, the new regulations required costly changes to be made in a relatively short time.

Not in my backyard

As old dumps were forced to close, many cities began searching for new landfill space that would meet EPA standards. Due to rising construction costs, some cities discovered it was less expensive to ship their garbage to other places than to upgrade old landfills or open new ones. Residents on the receiving end of shipped-in garbage were angered and even frightened, particularly when the waste contained hazardous substances. Protest movements sprang up in many places, the underlying rationale for which is often referred to as NIMBY, for "Not In My Back Yard."

Even though controls have tightened, shipping garbage (both hazardous and nonhazardous) to out-of-state destinations still goes on. States cannot pass laws prohibiting it because interstate commerce (trade across state lines) is protected by the Constitution.

A citizen protest in West Virginia in 1992 may have found a way around this obstacle. When a company in Philadelphia planned to open an enormous landfill in a rural area of West Virginia, residents living near the proposed site rallied in protest. A 1992 article in *E Magazine* says, "Capels Resources of Philadelphia thought one of the nation's poorest areas would be ideal for a giant landfill able to take 182 railroad cars a day of East Coast garbage. Local politicians drooled over the projected 367 jobs and the $6 to $8 million a year from taxing the trash. But residents rebelled."

What the residents did was appeal to the state legislature to stop the landfill. Knowing they could not interfere with interstate commerce, the lawmakers passed legislation limiting the amount of garbage that could be dumped in state landfills per month. The standard they set was just enough to meet their own state and local needs, but far below the amount needed for the proposed landfill. As one of the protest leaders put it, "West Virginia's new law doesn't ban out-of-state waste, it just favors landfills for local needs."

According to *E Magazine*, "Alabama and West Virginia have the two strictest laws on out-of-state waste in the country. If they hold up in court, the profile of the country's waste business could change. Large cities may have to keep their trash home, rather than trucking it to rural states."

Demonstrators protest the dumping of contaminated waste. Hazardous waste is sometimes shipped to poor, rural areas for disposal—a practice which local residents resist.

"THE TOXIC WASTE DUMP MEANS A LOT OF JOBS FOR THE AREA... DOCTORS, NURSES, HEALTH THERAPISTS..."

In another protest action in 1992, residents of a rural area in the Adirondack Mountains of New York organized to defeat a proposed landfill to bury ash from a neighboring county's incinerator. Not expecting serious opposition, consultants hired by the county chose 375 acres for the new landfill—forty miles from the incinerator.

Residents mounted such a vigorous protest that the plan was dropped. Environmentalist Bill McKibben, a resident of the area and one of the protesters, argues that the landfill was unnecessary in the first place. Writing in *Audubon*, he comments:

> It turned out that once the regional agency rejected the landfill, the county found out it needed a lot less space—10 or 20

> acres maybe. And the county next door was already planning to build its own oversize monstrosity of a dump, and it turned out they were desperate to rent space in it for a fraction of what it would have cost to build our own.

Facing increasing opposition from the public, along with rising construction costs, many small cities and towns are joining together to create regional landfills serving a number of adjoining areas. According to *Governing*:

> The Eastern Plains Council of Governments, based in Clovis, New Mexico, has launched an ambitious regional landfill planning effort that could be one of the most successful in the country. Working with 21 municipal and eight county governments, as well as a major U.S. Air Force base, the council is designing plans for a 320-acre landfill and a trash collection system with rural dropoff sites.

Innovations in landfill design and operation

Mandatory changes, such as the EPA's new landfill regulations, sometimes inspire imaginative solutions to old problems. That may be the case today as citizens, politicians, government officials, engineers, scientists, and planners join forces to meet the new guidelines.

In recent experiments, scientists have discovered a constructive use for leachate. Enclosing garbage in sanitary landfills slows down the natural process of decay, but circulating leachate through the garbage greatly speeds up decomposition. In a 1994 *Science News* article, environmental engineer Frederick G. Pohland writes, "Normally, a landfill takes 20 to 30 years to decompose. But using this method, it may only take 2 to 3 years. . . . We don't even have to add bacteria or chemicals. The landfill is biologically active and nutrient rich. This process simply accelerates natural decomposition." Leachate recirculation is being tested in experimental landfills called bioreactors at twenty sites around the country.

Methane gas is being recovered and sold as fuel at some landfills instead of being burned off at the site. An actual project of this sort is located in Whittier, California, at a landfill operated by the city of Los Angeles. Several million cubic feet of methane gas are recovered daily and sold to a California power company. In another California project, landfill methane is sold to an almond growers cooperative in Sacramento. It is then burned with almond shells to produce steam and electricity for industrial uses.

As suitable landfill sites become harder to find, and the rules for operating them more strict, an interesting question has been posed by some environmental engineers. Why not recycle entire landfills so the same ones may be used over and over again? Although such systems would be very costly to begin with, they would pay for themselves in the long run, according to proponents.

Environmental engineer Robert Landreth believes landfill recycling may be the way to go in the future. In *Science News,* Landreth says, "Fill one up, move to the next. Fill that one up, then move to a third. Keep going around, filling up sites and pumping out gas until the first one is fully cooked. Then dig up the first one, mine it for materials, and fill it again. You can go in a circle."

Landfill reclamation: new uses for old dumps

About twenty years ago, there were approximately twenty thousand landfills and garbage dumps operating throughout the nation, according to EPA estimates. Today, fewer than six thousand are still in use, and at least half of that number are expected to close in the near future. As troublesome as the new closure laws may be, they have proven beneficial in unexpected ways. Not only is the public health protected, but old landfills, once considered worthless, are being put to a variety of new uses.

Capped landfills are especially attractive to golf course developers because large tracts of land in urban areas are increasingly scarce or too costly for golf courses. Admittedly, in landfill golf courses, buried junk sometimes works its way out of the ground, and putting greens tip and sink as the garbage beneath them settles. Nevertheless, a 1995 article in the *Wall Street Journal* reports, "Many golfers love them. . . . For all the problems, there is still a powerful urge to turn wastelands into fairways."

An unusual recreational use for an old capped landfill is reported in the *Conservationist* of February 1993. "A model airplane club will rent the 17-acre landfill for use as an airfield," the article states. "Members of the LeRay Recycled Fliers will maintain the soil and vegetation and pay $1 a year in exchange for using the area to fly radio-controlled model airplanes."

Support exists for turning capped dumps into natural areas by replanting them with native trees, shrubs, grasses, and flowers. An organization known as Global ReLeaf specializes in this type of reclamation. In the Autumn 1995 *American Forests*, landscape architect Bill Young describes how he and his colleagues planted trees, shrubs, grasses, and wildflowers on a test plot at Fresh Kills, a twenty-four-hundred-acre dump site on Staten Island, New York, in 1989.

A similar project at an old landfill at Croton Point, New York, designed to attract birds and wildlife is described in a 1995 article in *Audubon*:

A test plot of the Fresh Kills capped landfill, seen here behind the housing development, was planted with flowers and vegetation. Old landfill sites can be replanted and used for recreational purposes.

> More than 1,000 pounds of seeds, including wheat, rye, dwarf corn, and wild broccoli, were planted to feed and shelter migrating birds and provide habitat for other wildlife. . . . By spring, the capped mound of garbage was rippling with tall grasses and dotted with daisies, blue bachelor buttons, and red clover. . . . Red-winged blackbirds flutter in and out of the grass, and ospreys sail above.

The outlook for landfills

At present, about 63 percent of all solid waste ends up in landfills (based on figures in *Statistical Abstracts of the United States, 1995*), a trend that will not be quickly or easily reversed. New ideas being tested today will take time to put into practice. Moreover, waste companies and municipalities that have invested millions of dollars in landfill facilities and equipment will be understandably reluctant to abandon existing systems.

Controversies over landfill location will undoubtedly continue for a long time, too. Most people accept landfills as a necessary if unpleasant part of modern life. Unless, of course, they get too close to home. As the commission that studied the Scratch Gravel Landfill in Montana in 1973 concluded, "Once the garbage has been thrown away people want it to disappear; they have no interest in where it ends up as long as it is not near them."

Today, in one way or another, garbage is near all of us. Even if it is out of sight, citizens are paying heavily for its disposal in fees and taxes. Waste collection and landfill operations consume sizeable chunks of most city budgets. Seeking relief from overflowing landfills, many municipalities are turning to an alternative method of garbage disposal—incineration.

2

Burning Garbage in Incinerators

DISPOSAL OF GARBAGE by burning is a practice as old as mankind. Prehistoric people tossed food scraps into their campfires, and, until recently, American families burned their garbage in backyard trash burners. Today, open burning has been banned in most communities, but a large number of municipalities have constructed gigantic incinerators to help dispose of wastes.

History of incineration in America

Incinerators were first invented and used in Europe in the 1870s. Just a few years later, they began appearing in major U.S. cities such as New York City. For years, that city's garbage had been dumped into the ocean or discarded in open landfills. But the refuse in the ocean had begun washing back onto the beaches, and dumps were fouling the air and spreading disease.

In 1897 George Waring, newly appointed chief of the New York City Sanitation Department, began a massive cleanup that included the construction of incinerators. Workers removed marketable items from incoming trash, and the remainder was burned in brick furnaces, generating temperatures up to 2000° F.

In 1905 another incinerator in New York City began using heat from burning garbage to produce steam. The steam was then used to generate electricity, which was sold to light a nearby bridge. Using garbage to produce energy for sale didn't catch on, however. In the early twentieth century, it was cheaper to use coal for that purpose.

Incinerators continued to operate in a few cities, but they were not overburdened. Dumping was still the main way to get rid of garbage. Furthermore, most people routinely (and legally) burned at least some of their own trash and yard wastes in backyard incinerators, even in large cities. According to Louis Blumberg and Robert Gottlieb, authors of *War On Waste*, more than one and a half million backyard incinerators were in use throughout Los Angeles County in the 1950s.

Air quality laws were finally passed in the 1970s, not only in Los Angeles but all over America, which limited backyard burning or banned it altogether. Air quality improved as a result, but tons of extra garbage were added to the landfills. With a landfill crisis looming, hard-pressed sanitation officials began to think about bringing incinerators back.

The big drawback, though, was financing them. Modern incinerators are gigantic plants that cover acres of ground and cost millions of dollars to build and operate. In addi-

A view of a modern incinerator. Sanitation problems in New York in the late 1800s prompted the development of incinerators, but they were not considered for widescale use until the 1970s.

tion to furnaces, they must be equipped with extensive pollution control devices to collect harmful gases and ashes. If they are to be used as waste-to-energy plants, boilers and turbines to create electricity from steam must be installed. Once constructed, their operation requires a highly trained staff, including engineers with specialized skills in mechanics, pollution control, and sanitation.

Not even larger cities could afford to build incinerators by themselves, but they could join forces with banking institutions and private industries to help finance, build, and operate the plants. Naturally, there had to be some incentive for the financiers—in this case, the chance to make a lot of money. With garbage piling up and landfill space running out, that chance seemed very good. Industrial companies, bankers, and investors were soon making deals to build multi-million-dollar incinerators in many of America's large cities.

Types of incinerators

Although there are many variations in design and construction, most incinerators built in the United States today are one of two types: refuse derived fuel systems (RDF) or mass burn furnaces.

In RDF systems, unburnable items are first sorted out from the incoming garbage. An article in the *Encyclopedia of Energy Technology and the Environment* explains: "Collection vehicles discharge waste loads onto the tipping floor [large dumping area]. This allows visual inspection of the MSW [municipal solid waste] and removal of bulky or oversized materials that may not be processable in the system or could be a potential hazard during waste processing."

The remaining garbage is then loaded onto conveyor belts for further separation. Shakers remove small particles that might harm the equipment, such as grit and sand. Magnets remove certain kinds of metal. The rest of the garbage is shredded in giant machines and compressed into pellets which are sold as fuel or used to generate electricity at the RDF plant itself.

A crane picks up a load of garbage from the pit of an incinerator. Mass burn facilities remove bulky, unburnable, or potentially harmful items from the pit before incinerating the rest of the waste.

The majority of incinerators in the United States today, however, are mass burn facilities, about 80 percent of which are also waste-to-energy plants (WTE), using garbage to generate electricity. As the name suggests, garbage handled in mass burn furnaces is neither separated nor shredded prior to burning. Almost everything is destroyed with the exception of bulky unburnable items (kitchen appliances, for instance).

Municipal solid waste fed into mass burn furnaces includes glass, wood, plastics, paper, food wastes, yard and garden debris, textiles, rubber, chemicals, solvents, ferrous metals (containing iron), and some nonferrous metals such as aluminum. The waste is first dumped into a large pit inside the plant where overhead cranes distribute it evenly and remove unburnable or potentially dangerous materials. From the pit, garbage is fed into chutes that convey it to the furnace.

Several types of mass burn furnaces have been developed by different companies; all must meet rigid EPA standards and be equipped with pollution control devices and ash removal systems. In addition, new incinerators must pass a test burn before being issued a permit to operate. Plants that burn hazardous wastes must meet even stricter standards.

A large number of small incinerators based on mass burn mechanics are operated by numerous medical facilities today to destroy medical waste. Though small, medical incinerators must also meet EPA standards. In 1988, after large quantities of medical waste washed up on Atlantic beaches, disposal standards were upgraded and now fall under the jurisdiction of the Resource Conservation and Recovery Act.

Although the new rules allow alternative methods of medical waste disposal, incineration is encouraged. An EPA publication states:

> The onsite incineration of medical waste has many advantages. Incineration sterilizes pathogenic [disease-causing] wastes; provides volume and mass reductions of up to 90 to 95 percent; converts offensive waste, such as animal carcasses, to innocuous [harmless] waste; can provide waste-heat recovery; and in some cases, can be used simultaneously to dispose of hazardous chemicals and low-level radioactive waste.

Problems with incinerators

Mass burn incinerators built in the United States in the 1970s and 1980s were patterned after systems operating in Europe, where the technology originated. Unfortunately, European incinerator designs did not work well with "American" garbage. Not only do Americans produce a much greater volume of garbage than do Europeans, but American garbage tends to be of a different type. The authors of *Rush to Burn* explain: "Because it contains more plastics and includes products that Europeans recycle, American trash tends to produce acid gasses that add to pollution problems and corrode plant equipment, requiring time-consuming repairs."

Economics also plays a part. "In this country, more than in Europe, resource recovery [producing energy from garbage] is a money-making enterprise," the authors of *Rush to Burn* continue. "Energy sales account for more than half the revenues of most large plants, and their operators are under great pressure to turn profits. And so, American plants are generally bigger and their steam boilers run hotter than their European counterparts, in order to sell more electricity."

The fledgling technology often led to costly mistakes by waste management companies. The authors of *Rush to Burn* describe one expensive incinerator breakdown in Florida:

> Trouble started for the $160 million Pinellas plant soon after it opened in 1983, as the difficulty in producing electricity from American garbage became apparent. Boiler tubes and the superheater that push steam through the plant's electrical turbines began to erode and break. The high concentration of plastics in the garbage, combined with the high temperatures,

Garbage is dumped into a chute that conveys it to the furnace. Waste-to-energy incinerators use the heat of the burning garbage to produce steam, which in turn produces electricity.

> caused acid gasses to build up in the Pinellas boiler. Repairs cost $8 million. The unscheduled shutdowns caused a buildup of 25,000 tons of garbage, which cost another $780,000 to cart away.

While advocates of incinerators concede numerous obstacles to overcome, they point out that many incineration plants in the United States operate successfully. As for those that do have problems, supporters insist that a certain amount of trial and error is unavoidable when putting a new technology into practice. Public fears are natural, too, as an EPA administrator states in a 1993 issue of *Chemical & Engineering News*:

> I think folks are obviously concerned about their health and concerned about their kids. Any new type of technology that may not be fully familiar, that may introduce some kinds of toxics into the air, causes them to be concerned . . . [but] I think we recognize that as technologies go right now, incineration destroys much more than some of the other technologies. And destruction is an attractive feature as opposed to piling [the wastes] up and capping it with dirt or somehow entombing it.

Disposing of the ashes

Incinerators, however, do not end the need for landfills. On the average, burning reduces the volume of garbage by 70 to 90 percent (depending on the type of garbage being burned) and reduces the weight by 60 to 75 percent. The remaining ash must be disposed of in landfills.

Incinerators produce two types of ash, fly ash and bottom ash. Fly ash particles are collected from the air pollution control system and are heavily contaminated with pollutants. Bottom ash is the residue that falls to the bottom of an incinerator during burning. It also contains hazardous substances, but is not as toxic as fly ash. Because the ash in municipal incinerators comes from household garbage, the EPA initially classified it as nonhazardous waste. This ruling allowed plant operators to bury the ash in municipal landfills instead of taking it to more expensive hazardous waste landfills.

Environmentalists were alarmed, charging that the EPA's position was not only unsafe, but unscientific. "Incineration differs from other forms of waste management in that major portions of the waste stream are physically and chemically transformed during the combustion process," Richard A. Denison and John Ruston, editors of *Recycling and Incineration*, explain. "Moreover, the products of this process—both solids and gasses—differ markedly from the original waste in their environmental and biological behavior."

Nevertheless, the EPA left it up to plant operators to decide whether the ash was hazardous, with no tests or proofs required. It became common practice to mix the more toxic fly ash with the bottom ash and dispose of it in municipal solid waste landfills. Ash was also sold to contractors for road building and other construction purposes.

In an effort to stop these practices, the Environmental Defense Fund, a nonprofit environmental group, sued the city of Chicago over unsafe disposal of waste from its incinerators. The dispute

The manager of an incineration plant opens a window to show trash burning in the furnace. Although incineration reduces the volume of waste, it leaves behind hazardous ash that needs to be disposed of.

eventually came before the U.S. Supreme Court in 1994. A report of the Court's decision in *Chemical & Engineering News* states:

> The Supreme Court has ruled 7 to 2 that ash from municipal incinerators must be treated as hazardous waste. . . . The Court sided with the Environmental Defense Fund, which sued to have the waste declared hazardous because it may contain levels of cadmium and lead that exceed federal waste standards. Under the Court ruling, such waste will have to be disposed of in hazardous waste landfills.

For the incinerator operators, and for the public as well, it was a costly decision. The increased expense of using hazardous landfills will eventually be passed on to city residents in higher garbage collection fees. But in terms of public health and safety, it was a major victory.

Smokestack emissions

Besides ash, incinerators produce other potentially harmful emissions. Plants are required to meet EPA clean air emission standards, but there is wide disagreement about what these standards should be or how to measure them. A 1993 article in *Chemical & Engineering News*, for example, tells of an incinerator in Arkansas that was temporarily shut down on a complaint filed by the environmental organization Greenpeace. Greenpeace engineers calculated that only 99.96 percent of dioxin (a toxic by-product of incineration) was being destroyed instead of 99.9999 percent, as required by the EPA. The incinerator operators said Greenpeace engineers were basing their calculations on the wrong data. Their own figures proved the regulations were being met.

For the average citizen, unequipped to evaluate such technical arguments, it eventually comes down to whom one wishes to believe. Even the experts don't always know how much of a toxic gas released into the air is safe, if indeed any amount is safe. And the critical issue of the long-range effects of incinerator gases on human health is yet undetermined.

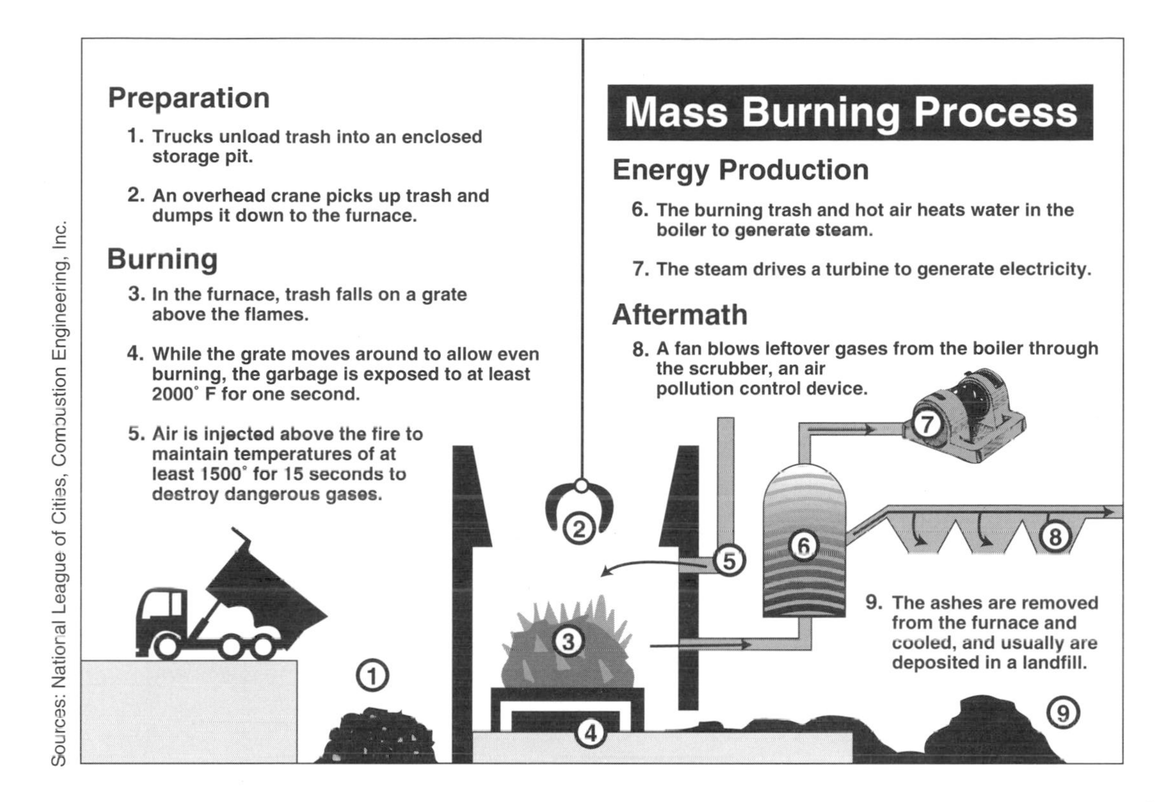

Sources: National League of Cities, Combustion Engineering, Inc.

Independent health studies (that is, conducted by those who have no interest in the outcome) suggest that a link between incinerators and certain diseases may exist. A 1993 issue of *Science News* describes several recent studies, all showing a greater incidence of health problems (including nervous system ailments, memory loss, and sleep disorders) in people living near incinerators (or working at them) than in a control sample who lived and worked farther away. However, the article cautiously concludes, "None of these studies proves that incinerators harm health. But they do raise strong suspicions that the apparent links are real."

Increased risk of cancer is another fear associated with incinerators. In Kentucky, a grassroots organization called Coalition for Health Concern was founded by Corrine Whitehead after she became convinced that a hazardous waste incinerator was causing a high rate of cancer in her community.

The incinerator company, LWD, Inc., insists it is not at fault. A 1993 article in *New Choices* states:

> Officials at LWD, Inc. . . . refute the charge that their facility causes cancer. . . . They cite recent reports from state environmental authorities that found no acute toxic problems with the air in and around Calvert City [Kentucky] and no substantial threat to public health or the environment for the possible release of hazardous substances at the incinerator site.

Reports from state environmental authorities do not always ease people's minds, however. In *Rush to Burn*, the authors criticize the actions of two state officials whose job it was to test for hazardous emissions from a garbage incinerator in Maryland: "They stood a few yards from the . . . incinerator, glancing from their watches to the smokestack every 15 seconds. The sky seemed smoke-free. . . . So, after 15 minutes they drove back downtown and wrote a report that recommended a one-year renewal of the plant's license to burn."

In whose backyard?

On the whole, public response to incinerators has not been enthusiastic. Nevertheless, some communities tolerate them because they bring jobs to economically depressed areas. All too often, however, the plants are located in those areas simply because the residents are powerless to stop them. Successful NIMBY protests typically occur in neighborhoods where residents are better equipped to fight legal battles against the promoters of incinerator projects.

While NIMBY protests are understandable, chances are incinerators are going to be built in somebody's backyard. That somebody usually turns out to be the poor and powerless, who tolerate a disproportionate share of other hazardous sites as well as incinerators. A 1996 article in the *ZPG (Zero Population Growth) Reporter* states, "Low-income communities host a large share of the nation's toxic waste dumps, landfills and hazardous industries. People of color have the highest rates of pollution-related diseases. These groups also have less access to health care and have fewer economic and political resources with which to address their grievances and problems."

Mobilized by this imbalance, a rising environmental justice movement is calling for reform. "Political action groups in the United States are increasingly concerned about the fairness with which environmental risks are geographically distributed," a 1993 article in *Environment* reports. "The efforts of these groups—from environmental organizations to advocates for the rights of minorities—are affecting neighborhood community organizations, state and federal legislative initiatives, and international environmental agreements."

An example of the growing power of minority protest groups occurred in an inner-city neighborhood of Los Angeles, California, in 1980 when residents rallied to defeat a mass burn facility slated for construction in their area. Believing incinerators were the answer to the city's overflowing landfills, Los Angeles city officials planned to erect three mass burn facilities in different sections of the city, to be built one at a time.

"I think we're making progress . . . They printed their latest incineration proposal on recycled paper."

Based on the advice of consultants hired by the city, the site chosen for the first incinerator was a neighborhood in South Central Los Angeles, a community Blumberg and Gottlieb describe as "young, poor, and heavily minority. . . . 40 percent of the residents had incomes below the poverty level. . . . The population was 52 percent black and 44 percent Hispanic."

The protest launched by residents of that neighborhood was quickly joined by civic and environmental organizations from all over the city. The slogan changed from "not in my backyard" to "not in anybody's backyard." Under extreme pressure, the mayor of Los Angeles withdrew his support for the incinerator in 1987, and the city council officially killed the project shortly afterward.

At about the same time, protesters in East Liverpool, Ohio, were unable to stop the construction of an incinerator in their small, economically depressed community. However, public demonstrations and lawsuits forced it to remain closed for several years. Builders of the hazardous waste burning plant emphasized its economic advantages to the community and insisted that the majority of East Liverpool's citizens were in favor of it. Opponents claimed their community was being used for gain by powerful business interests with little regard for East Liverpool's citizens. Over vigorous protest, the plant finally opened in 1993 and continues to operate today.

Public opposition like that in Los Angeles and East Liverpool slowed the fast-growing incineration industry for a while, but, for lack of other choices, it seems to be making a slow comeback.

Revival of incineration

The *Encyclopedia of Energy Technology and the Environment* breaks down current statistics on functioning incinerators as follows: 168 municipal solid waste incinerators burning 26 million metric tons per year; 171 licensed hazardous waste incinerators burning 1.3 million metric tons per year; 6,850 medical waste incinerators disposing of 400,000 metric tons of medical waste per year.

Today, about 16 percent of all waste generated in the United States is incinerated (based on figures in *Statistical Abstracts of the United States, 1995*). With several new plants currently under construction and others in the planning stage, that figure is expected to rise. If incineration does make a comeback, it will be attended by much stricter controls than ever before, which may be of some comfort to its critics. According to a 1994 news brief in *Chemical & Engineering News*, "EPA has taken 23 more enforcement actions against the owners and operators of incinerators, boilers, and industrial furnaces that burn hazardous wastes. The actions are part of EPA's continuing enforcement initiative on combustion that began last year."

Protesters in East Liverpool, Ohio, attempt to stop the opening of an incinerator built in their community. Despite the residents' opposition, the plant was opened in 1993.

Incinerators and recycling

When opponents of incinerators and landfills are asked how they would handle the garbage problem, they invariably reply, "Recycle, reuse, reduce!" They argue that incinerators, particularly the mass burn types, have undermined recycling efforts. Waste-to-energy plants must be guaranteed a certain amount of garbage per month by municipalities to keep the plants operating and show a profit. When tons of waste are eliminated from the waste stream through recycling and composting, the required volume of garbage cannot always be supplied. Therefore, municipal governments that utilize incinerators may not always be eager to push recycling programs.

Furthermore, critics charge, the big money, both public funds and private capital, has gone into expensive landfill and incineration ventures, while recycling projects and the development of markets for recycled goods have been neglected.

3

Recycling Garbage

AMERICANS "TIGHTENED THEIR BELTS" and did without a lot of things throughout the Great Depression and during World War II. They conserved natural resources and they reused and recycled clothing, newspapers, tires, tin cans, tools, household goods, and many other items. After World War II, this spirit of conservation was rather quickly cast aside. Within a few decades, America became the most wasteful nation on earth in terms of garbage generated per person. In *War on Waste*, a public official offers an explanation for this transformation:

> How did it happen that we went from learning to conserve to learning to be wasteful? During the economic and international crises of the 1930s and 1940s, policy-makers saw no choice but to teach conservation. . . . But when the end of the war brought the possibility of yet another depression, policy-makers essentially panicked. They no longer wanted us to conserve; they wanted us to buy, buy, buy. . . . Our production-oriented society changed to a consumption-oriented society. When it comes to resource policy, we have lost the conservation and recycling worldview that used to be a part of everyday life.

Today, many Americans are hopeful that these values can be revived. But to accomplish this, more citizens will have to participate in community recycling programs, and industry and business will have to change some of their current wasteful practices. Even more importantly, markets for both waste materials and recycled goods will have to increase.

Recycling falters

The importance of markets for waste and recycled products was not clearly grasped in the early days of environmental awareness. Following the first Earth Day in 1970, community recycling programs sprang up all over America. Thousands of citizens willingly participated in these programs, whose primary goal was to conserve the earth's dwindling natural resources.

In hundreds of communities, recycling centers were opened by civic and environmental groups to which residents brought their household waste such as newspapers, cardboard boxes, glass bottles and jars, steel and aluminum cans, and certain types of plastics. From the collection center, waste materials were sold to processing companies at very low prices—if they could be sold at all.

Lack of markets for collected wastes soon made it clear that collection is only one part of the process. Recycling has three phases, symbolized by a loop of three chasing arrows often printed on recycled products. In phase one, waste is collected and processed into raw materials (old newspapers are turned into paper pulp, for example). In phase two, products are manufactured from recycled raw materials (paper pulp is transformed into rolls of newsprint). In the third phase, recycled products

People flocked to recycling bins like these after Earth Day in 1970 sparked environmental awareness; however, the recyclables lacked a market.

are purchased and used by consumers (citizens buy daily newspapers printed on recycled newsprint). Once the loop has been set in motion, it has no beginning or end. All phases blend into one another in a continuous stream.

Throughout the 1970s and 1980s, collection of recyclable wastes outdistanced the ability of processors and manufacturers of recycled goods to keep up with it. As a result, many communities soon found themselves with vast amounts of waste materials on hand that nobody wanted. In a 1995 article in *U.S. News & World Report*, staff reporter David Fischer writes:

> Between 1988 and 1992, the number of nationwide curbside recycling programs more than quadrupled, from 1,042 to over 5,400, because of the burgeoning [growing] environmental movement and a perceived landfill shortage. But the production of waste processing facilities and factories with the technologies necessary to turn yesterday's garbage into tomorrow's paper, cans and bottles did not keep pace. As a result, prices of recyclables plummeted. The glut of these materials was so great that between 1990 and 1993 some U.S. cities even had to pay recyclers to take old newspapers off their hands.

Local social and civic groups proved successful at organizing community recycling projects, but they were not in a position to do much about markets for selling the collected materials. That part of the loop needed assistance from government and industry that was slow in coming. Most local governments had invested heavily in landfills and incinerators to manage their waste disposal problems, and were not enthusiastic about recycling. Not all industrialists were pleased with recycling either, considering the costs of converting plants to new manufacturing procedures.

Revival of recycling

In spite of obstacles and setbacks, believers in recycling did not give up. Their determination paid off beginning around 1994 when old newspapers, clear glass, aluminum cans, and plastic bottles began bringing much higher prices. "Since May, 1993, the price paid by processors for used, clear glass containers has jumped 78 percent, as has

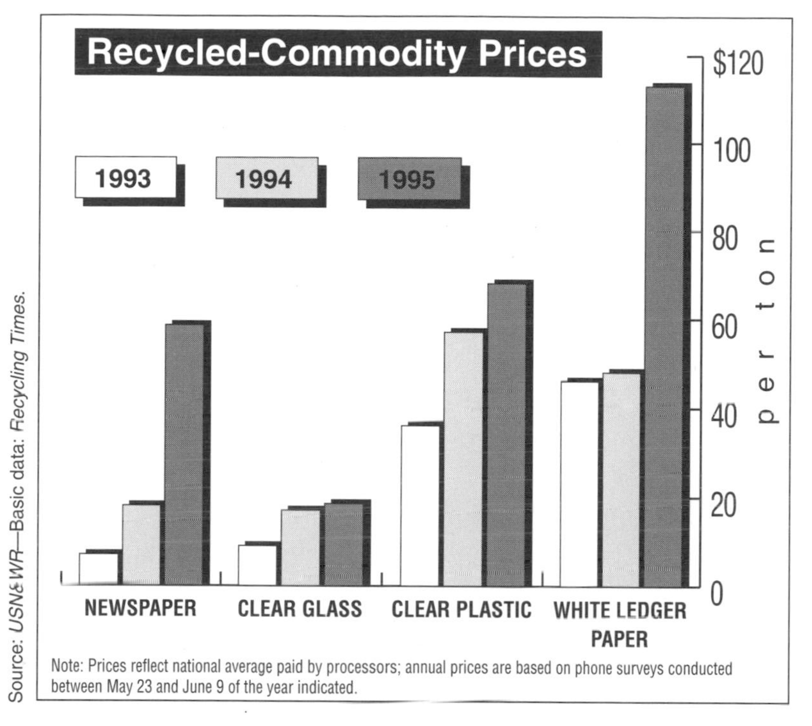

the value of discarded aluminum cans," Fischer states in *U.S. News & World Report*. "Over the same period, the cost of old newspapers has skyrocketed 1,338 percent." In 1995 *Business Week* reported, "Demand for polyethylene-terephthalate (PET) containers [clear plastic] . . . is growing by 21% a year, and prices have more than tripled since last November."

This seemingly miraculous turnaround in just a few years is the fortunate result of many social, political, and economic factors that started coming together in the mid-1990s. A wide variety of participants—environmental and civic organizations, government, business, industry, and millions of individual citizens—were instrumental in bringing about the change in attitude.

Support for recycling

Although many Americans were apathetic about garbage (beyond its effect on their personal lives), others joined the environmental movement that began in the 1970s and stubbornly pressured governments at all levels for passage of recycling laws and other environmental measures. Various organizations conducted educational campaigns, lobbied Congress, filed lawsuits, and protested the building of landfills and incinerators that destroyed tons of recyclable materials.

Although slow to support recycling in the beginning, government assistance also contributed to its comeback. Most states have created special recycling agencies to help local governments find outlets for their waste products and to assist companies in developing recycled products. Many states also provide tax breaks, grants, low-interest loans, and other incentives to companies involved in various aspects of recycling.

Another important source of government support is its purchase of recycled supplies for public offices and workshops. In a 1995 article in *American City & County*, editor Michael Terrazas states, "More and more local governments are making the commitment to close the recycling loop and buy recycled products. This, along with increased purchasing from the private sector, is prompting businesses to retool their production facilities to make recycled products."

Some cities have gone out of their way to give recycled products a chance. For example, city officials in Philadel-

A display of products made from recycled materials. One way in which the government has responded to citizens' environmental concerns is by buying recycled products for use in its offices.

phia decided to buy pencils made from recycled newspapers even though they cost 25 percent more than pencils made from wood. "As it turned out," Terrazas writes, "Philadelphia's purchase was large enough that it actually pushed the price of the [recycled pencil] down some, and after more large purchases from other governments and businesses across the country, the pencil now costs less than those made from virgin wood."

To deal with tons of old newspapers collected in the late 1980s, several states passed "recycled content laws" requiring publishers to use a certain amount of recycled paper in their publications. However, recycled content laws irritated many businesspeople who felt the passage of these laws interfered with free trade.

Popular or not, the laws had the desired effect. In the January 1995 issue of *Governing*, Tom Arrandale writes, "The paper industry . . . was reluctant to build expensive 'deinking' plants [ink removal] before they were convinced that publishers were willing to print on recycled paper. In 1988, there were just nine mills in North America equipped to deink newspapers and convert the fiber into more newsprint." After recycled content laws were passed, this situation changed dramatically.

"The industry has responded," Arrandale says, "by opening 20 new deinking plants in the past six years, and the demand for materials to keep those mills running has ended the glut of unmarketable newspapers that bedeviled recycling programs just a few years ago."

Voluntary recycling in business and industry

Businesses and industries also played a major role in the recycling comeback by initiating a wide variety of programs and projects on their own. For instance, the National Recycling Coalition (NRC), a nonprofit foundation, launched the Buy Recycled Alliance for business groups in 1992. Starting with only twenty-five companies, the alliance quickly grew to over five hundred members, all agreeing to purchase recycled materials and products whenever possible. Many businesses are discovering that

dedicated commitment to recycling can save money. American Airlines, for example, saved more than $100,000 in a year's time by purchasing recycled computer paper.

Moreover, many businesses are responding to the growing public demand for recycled products in order to gain or keep customer goodwill. Rubbermaid Commercial Products, Inc., maker of household items such as laundry baskets, kitchen and bath utensils, and storage boxes, is a major user of recycled HDPE plastics (high density polyethylene, used in milk jugs and detergent bottles). Rubbermaid pays more for the recycled plastic than it would have to pay for new plastic because it believes that's what customers want.

This bale of shredded recycled paper will be shipped to a mill, where it will be manufactured into new products.

Concern for the environment has led to other voluntary recycling projects. A 1995 article in *Nation's Business* describes the Walt Disney Company's contract with Global Green, Inc., of Georgia, to develop recyclable uniforms for their employees. "Disney was concerned because its uniforms are routinely disposed of in landfills," the article states, "and it was intrigued with the idea of fabrics that could be recycled and used again." The fabric developed by Global Green is made from recycled plastic bottles. At this point, it is somewhat more expensive than other fabrics, but Disney decided to use it for the sake of the environment.

In other volunteer actions, trade associations (groups of companies promoting a shared industry) assist the public with specialized recycling problems. The Society of the Plastics Industry, Inc., has developed a code for recycling various plastic materials that manufacturers stamp into finished products. The Steel Recycling Institute helps set up local recycling centers. The Glass Packaging Institute and Aluminum Association (among many others) offer educational materials for teachers.

Business opportunities in recycling

For many companies, recycling has become the main focus of their activities. "Business opportunities abound in municipal solid waste disposal," Susan Williams asserts in *Trash to Cash*. "The United States is dramatically transforming the way it handles that waste, providing openings for many new players to enter the field and for existing players to expand their services."

One type of new business that has benefited from the environmental movement is the so-called green company, offering recycled and environmentally safe products for the home and garden. Other business opportunities are opening up in the computer industry: Some companies salvage rapidly outdated machines for reusable parts and precious metals; others repair the machines and sell them at reduced prices. GreenDisk, a small company in Preston, Washington, recycles unopened boxes of out-of-date software. "From the 20 million pounds of boxed software it has processed," a 1995 article in *Science News* states, "GreenDisk has mined more than 10 million professional-grade disks."

Recycled glass presents still other business opportunities. Recycled glass crushed into a substance called cullet may be substituted for sand in asphalt paving material. The result is a sparkling surface called glasphalt that is durable and pleasing to the eye. "New York City is paving the way," *Trash to Cash* reports, "laying the sparkling glasphalt on more than 400 miles of streets, including Times Square, where the embedded glass reflects the flickering neon lights."

Unforeseen events gave plastic bottle recycling companies a boost in 1995. Due to a poor growing season, cotton crops in Asia were much reduced. As a consequence, Asian textile mills began using more virgin polyester, which was also in short supply. Prices for both cotton and polyester rose sharply, forcing American clothing manufacturers to raise their prices. According to a 1995 article in *Business Week*, this situation "drove textile producers into America's used-PET-bottle market where recycled containers are used to make polyester fibers that are then spun into everything from sweaters to upholstery. As a

These recycled glass bottles will be crushed and used in place of sand in asphalt. The so-called glasphalt makes a durable paved surface.

result, bottle prices shot from as low as 10 cents a pound a year ago to as much as 40 cents last spring."

PET and HDPE bottle recyclers are betting that this trend will continue, even when cotton and virgin polyester supplies rebound. Some are so confident, they are already expanding their facilities for future business. The major problem facing plastic bottle recyclers at present is lack of bottles. Half of the PET bottles produced in America still wind up in landfills or incinerators.

Today hundreds of businesses and industries in the United States are either partially or wholly based on recycling. While many of these organizations are large, well-known corporations, a great many others are small businesses taking a chance on the future of recycling.

Personal recycling

Countless small acts of families and individuals also contributed to the revival of recycling. Using both sides of writing paper, laundering cloth diapers, rinsing out plastic storage bags for reuse, bringing groceries home in canvas sacks, donating outgrown clothing to thrift shops, and separating recyclables from the trash all played a part.

Recyclers who created backyard compost piles from organic waste made a sizable contribution, too. The word *recycling* rarely calls to mind food scraps, leaves, grass, weeds, and tree limbs. Such materials can be turned into valuable fertilizers, however, by piling them in a heap or confining them in special containers until they decay. Composting not only creates useful fertilizers, it also removes tons of organic material from the municipal solid waste stream.

City officials in Chandler, Arizona, were able to solve two problems with composting. The population of Chandler (near Phoenix) has been growing rapidly over the past few years, with many new housing subdivisions under construction. With increased development came increased

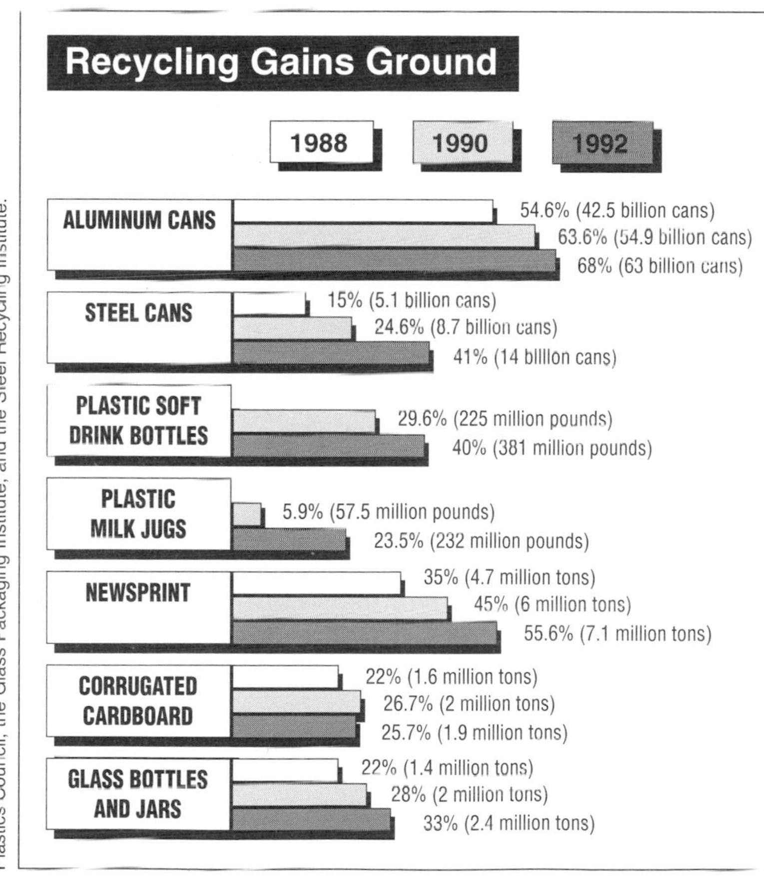

Sources: The Aluminum Association, the American Forest and Paper Association, the American Plastics Council, the Glass Packaging Institute, and the Steel Recycling Institute.

yard waste disposal problems. When the city was forced to replace hundreds of worn-out garbage containers made of unrecyclable plastic, the old containers were offered free of charge to Chandler's citizens for compost bins. The results have made the program worthwhile. "It is estimated that nearly 1,700 tons of yard trimmings have been source separated from the solid waste stream and made into a useful soil amendment," a 1996 article in *Biocycle* reports.

Problems in the private sector

The efforts of government, business, and the private sector have coalesced in very recent times to breathe new life into an ailing recycling movement. There are still, however, many problems to solve before recycling can make a significant difference in America's waste disposal crisis.

Inefficient collection of recyclables is one such problem. In a 1995 issue of *Waste Age*, Chaz Miller, manager of recycling programs for the Environmental Industry Association, says:

> Ideas for educating householders to be more efficient participants [in recycling programs] are abundant. These include getting neighbors to place their recyclables together in order to increase materials at stops, educating residents to set out full bags and bins instead of half-containers, collecting recyclables every other week, using larger collection containers, and asking participants to flatten plastic containers.

Each of these ideas has merit, but Miller believes economic incentives often work better. Some cities give rebates on garbage collection fees for households that participate in recycling programs. Other communities charge collection fees that rise as the amount of garbage that must be hauled away rises. "These systems have been shown time and again to result in increased amounts of materials set out for recycling," Miller says.

A view of plastic jugs left on the curb to be picked up and recycled. One problem with collecting recyclables is that they are not consolidated into large, compact, and efficient bundles.

Expenses for the industry

Picking up and transporting recyclables has become increasingly expensive for

Recycling trucks have changed to accommodate the larger amounts and more varied types of recyclables left on curbsides. Compartments on the side of this truck keep different materials separate from each other.

waste collection businesses. Specially designed trucks that are larger, more maneuverable, divided into compartments, side-loaded, and equipped with hydraulic lifters are necessary for efficient collection. Miller explains:

> In the mid-70s, curbside recycling was easy. Virtually all the programs collected only newspaper at the curbside. They were placed in a rack beneath the garbage truck or in a trailer behind the truck. . . . Today, curbside recycling is much more complicated. Curbside programs can collect a bewildering array of materials, including plastics, mixed paper, and even textiles.

Processors of waste materials have their problems, too. One is the condition of the material to be processed, whether the waste is glass, aluminum, newsprint, plastic, or steel. Newsprint, for example, must be deinked before it can be used. It also must have staples, glue, and other trash removed from it. Glass and plastic containers must be clean, and certain kinds of waste containing hazardous raw materials must be weeded out.

Another common problem is the immense cost of retooling old plants or setting up new ones to deal with recycled materials. A new deinking plant built by Union Camp Corporation in Franklin, Virginia, in 1994 cost $85 million. Added to the expense of facilities and machinery are the costs of research to develop recycled products.

As in any other business or industry, some recycling companies succeed and others fail. But despite the risks involved, new recycling plants continue to be built and older ones refitted. Since 1990 the number of companies that recycle plastic has tripled and eighty-five new paper mills producing recycled paper have been built.

Stabilizing the market

The recent upturn in waste markets is helping to complete the recycling loop, but recyclers are not complacent about the trend. Miller warns in his 1995 *Waste Age* article, "Bull markets [upswings] are not forever. Recyclers cannot afford to adopt a 'What, me worry?' attitude toward the business of recycling."

Prices for waste material will undoubtedly fluctuate as industries retool. An encouraging sign, however, is the opening of a recyclables exchange on the Chicago Board of Trade in 1995. The exchange will provide information on the types and quantities of waste material available to purchasers, thus helping establish more uniform prices. A 1995 article in the *Economist* explains: "The exchange will also establish some much-needed standards for the commodities traded on it. Used paper, for example, comes in versions as varied as newspapers and corrugated cardboard with differing degrees of contamination."

Recycling yesterday's wastes

Although systems like a recyclables exchange may help to quickly transfer garbage to eager recycling industries, some waste problems today are greatly complicated by vast accumulations from the past. One such problem is that of discarded tires. In addition to the 240 million tires being thrown away each year, there are 3 billion old tires lying in landfills, stockpiles, and illegal dumps. A 1994 article in *Popular Science* offers this perspective: "If you could stack them, the tires would stretch 142,000 miles high—more than halfway to the moon."

During World War II, old tires were retreaded and reused. After the war, inventions such as synthetic rubber and steel belting made tires less suitable for retreading. The result today is mountains of old tires all over the country that take up great amounts of landfill space or are stored in unsightly and dangerous stockpiles. In 1983 thousands of tires stored near Winchester, Virginia, caught fire and burned for nine months. The smoldering mass polluted air and water sources and cost millions of dollars to clean up.

Not for lack of effort are the mountains growing. Some industries are burning used tires in waste-to-energy plants. Paving made from chopped-up tires is being used for road construction. In some places, the walls of energy-efficient houses are constructed from old tires packed with dirt. Nevertheless, as beneficial as these measures are, they don't even make a dent in the staggering piles of discarded tires.

Scientists and engineers believe the best solution is to break the tires down into their original components—oil,

A glut of old tires has accumulated since people stopped re-treading and reusing them about fifty years ago. Heaps of tires, such as this one, pose possible health risks to people in the vicinity.

rubber, steel, etc.—all of which can then be recycled. However, the high temperature needed to melt tires unfortunately destroys much of their raw materials. That problem may have been solved by a recent invention. According to a news brief in a 1993 issue of *Business Week*, "Titan Technologies Inc. in Albuquerque says a secret brew of catalysts [reaction agents] heated to 450°F breaks down tires into their components—oil, carbon black, and steel—which can then be reused." Two tire recycling plants based on this technology have already been built in Korea.

As a rule, however, complex problems are rarely solved by one major breakthrough. As a Goodyear Rubber Company official says in *Popular Science*, "You can only make so many swings and roadways with old tires. We'll need a combination of many ideas."

An opposing view of recycling

Despite recycling's increasing popularity in America, not everyone thinks it is a wise course to follow, or even a necessary one. Many of those who question the wisdom of recycling are not so much against the idea of recycling as they are against governmental intervention to make it work. They feel that laws such as recycled content regulations and packaging requirements stifle the freedom of business and industry to create the best possible products for their customers. If left alone, they argue, business and industry will respond to environmental concerns voluntarily.

Critics of recycling believe mandated content laws which require manufacturers to recycle a certain percentage of their waste packaging into new packaging are not always the best way to deal with waste. Other factors also have to be considered, such as the energy needed to recycle many wastes. In a 1995 article in *Environment*, Christopher Boerner and Kenneth Chilton, associates of the Center for American Business, say, "In many circumstances, simply recovering the energy from packaging in a waste-to-energy incineration facility may be more eco-

nomically efficient than recycling. Mandated content requirements, however, prevent these and other options from being considered by manufacturers or local solid waste officials."

In addition to economic misconceptions, skeptics believe the recycling movement is based on a number of false premises, one of which is that landfill space is running out. This is simply not so, critics insist. Landfill siting problems are due more to public opposition and stricter regulations than to land scarcity. Certain widely publicized news stories have had an inhibiting effect, too, such as the account of the New York garbage scow Mobro, which could find no place to dump its load in 1978.

In a 1996 article in the *New York Times Magazine*, staff writer John Tierney writes:

> The Mobro's saga was presented as a grim harbinger [omen] of future landfill scarcity, but it actually represented a short-lived scare caused by new environmental regulations. . . . America today has a good deal more landfill space available than it did ten years ago. Landfills are scarce in just a few places, notably the Northeast, partly because of local economic realities (open land is expensive near cities) but mainly because of local politics.

In areas that have little landfill space, most communities ship their garbage to remote areas where landfill operators are often hungry for the business.

Boerner and Chilton agree that landfill scarcity is a myth, and add another misconception to the list—landfills and incinerators increase health risks in surrounding areas. They counter:

> Increasingly stringent construction and post-closure monitoring requirements . . . are reducing the already infinitesimal health risks associated with landfills. EPA estimates a risk of less than one cancer incident in 13 years as a result of all currently operating landfills in the United States. The agency goes on to point out that modern waste-to-energy plants pose a cancer threat of less than one in a million.

Probably the most widespread misconception, however, is that recycling itself has no negative impact on the environment. According to Boerner and Chilton, "Recycling . . . is a manufacturing process like any other. Raw materials must be collected, prepared for processing, and manufactured into marketable goods. Although the process of recycling conserves some resources, it consumes others."

Overall, critics see recycling as an activity that is simply not worth the time, money, and energy expended to accomplish it. The popularity of recycling, they maintain, is based on sentiment (people desperately wanting to make a difference) and mistaken ideas about a garbage crisis.

Workers pile up bags of aluminum cans to be recycled. Some people claim that recycling is not worthwhile since the process itself consumes resources and landfill space is not as scarce as it may seem.

According to Tierney, there is no garbage crisis. He says, "Recycling, which was originally justified as the only solution to a desperate national problem [landfill shortage], has become a goal in itself—a goal so important that we must preserve the original problem."

In spite of its critics, the recycling movement is growing stronger today. The longer it can maintain its momentum, the more likely it is to become a permanent part of American life. If it is to survive, both governments and individual citizens must not only give it a chance, but actively participate.

A government success story

In the literature on recycling, an often repeated theme is that recycling must be given an equal footing with incineration and landfilling if it is to survive and prosper. One American city that decided to do just that is Seattle, Washington. After closing its overflowing landfills in the mid-1980s, city officials in Seattle considered building an incinerator, and hired consultants to draw up an incineration plan.

As the study progressed, city officials became concerned about the effects an incinerator would have on the environment. They were also worried about the cost of hazardous ash disposal. Putting the incinerator plans on hold, the city council commissioned a study comparing incineration with large-scale recycling. They wanted to know how much recycling they could get for the cost of an incinerator.

The detailed study, called *Recycling Potential Assessment*, was completed in 1988. According to Denison and Ruston in *Recycling and Incineration*:

> The study estimated that recycling levels of 50 to 78 percent are feasible in Seattle, and that on a life-cycle, cost-per-ton basis, even very high rates of recycling would be less expensive than any level of incineration. After the study was concluded, the mayor and city council decided to table the incinerator, and instead set the city on a path of achieving 40 percent recycling by 1992, 50 percent by 1993, and 60 percent by 1998.

So far, the program has proven successful. An update of the *Recycling Potential Assessment* in May 1994 states, "Recycling has proven to be very cost effective. Between 1988 and 1993, Seattle's residents saved $8.5 million as a result of utility-sponsored recycling programs. In other words, if the city had not provided recycling or yard waste collection but had collected all those materials as garbage, it would have cost $8.5 million more."

According to Thomas M. Tierney, director of Seattle's Office of Management and Planning, the success of Seattle's recycling venture is largely due to educational and community outreach programs encouraging recycling and resource reduction. These include a paper reduction program for businesses, a backyard composting program in which thirty thousand bins have been distributed to citizens, "green" programs showing how to garden and clean with nontoxic products, a "smart shopping" program to reduce excess packaging, and a directory of where to buy used items and have broken articles repaired.

Recycling success has not made Seattle's city officials complacent, however. The 1994 *Recycling Potential Assessment* adds, "There are still significant quantities of recyclable material going to the landfill. . . . Not only are these items present in large quantities, they can also be readily recycled. Thus, they represent key targets for 'Phase II' of Seattle's recycling efforts."

While Seattle may be considered a model, many other communities are striving toward increased recycling as well. *Beyond 40 Percent* features sixteen additional cities of various sizes that have placed high priority on recycling.

An individual success story

Sometimes individuals lead the way. Steven Loken, a builder in Missoula, Montana, constructed his home completely from recycled materials. The house is very traditional in style, because, as Loken says in an article in *Parade*, March 3, 1996, "I wanted to show that you could use recycled building materials without making any compromises on the type of house most Americans want. This

These plastic bottles will be cut up and used to make carpets. Individuals as well as communities have become interested in conserving resources and reusing products.

meant the house had to look like any other house if the ideas behind it were going to catch on."

In his house, Loken used floor and bathroom tiles made from recycled car windshields and fluorescent lightbulbs, carpets woven from recovered fibers and plastic milk jugs, carpet pads made from used tires, wood siding salvaged from demolished buildings or molded out of lumber chips, and insulation derived from recycled newspapers. *Parade* reports that at least twelve thousand people from all over the world have come to see the house since Loken built it in 1990. It has also become the prototype for houses built in several other states.

To help spread his ideas, Loken founded an organization in Missoula called the Center for Resourceful Building Technology. "I'm interested in cutting down on waste altogether and making more intelligent use of our resources," he says. "If we could master this, our environment and our society would be a lot better off."

The future of recycling

Whether it is being practiced by an entire city or a dedicated individual, recycling is not just a simple process of collecting cans and newspapers to make citizens feel good about the environment. Of coure, feeling good may be a worthwhile by-product, but recycling is a business that must prove its value in the marketplace if it is to succeed. If given a fair chance, many Americans are convinced recycling can go a long way toward solving the nation's waste problems.

According to EPA statistics, progress toward that goal is being made. In 1993, 21.7 percent of all waste generated in the United States was recovered compared to 6.7 percent in 1960. The next task, and probably the most difficult of all, is convincing Americans not to make so much garbage in the first place.

4

Reducing Garbage at the Source

THE TITLES GIVEN to waste management acts passed by Congress over the past thirty years reveal how the scope of waste management has expanded. First was the Solid Waste Disposal Act of 1965, which emphasized the destruction of garbage. In 1970 the Resource Recovery Act added recycling to waste management goals. Six years later, the Resource Conservation and Recovery Act recognized source reduction—creating less garbage—as an equally important goal of waste management.

Source reduction is less an activity in itself than the result of other activities. When plastic grocery bags are reused and recycled, for instance, fewer new bags are needed, thus saving virgin raw materials (oil and chemicals). Similarly, making newsprint from old newspapers conserves trees from which most new paper is produced, and recycling aluminum cans results in less demand on aluminum ore deposits.

At the consumer level, source reduction involves disciplining oneself to shop more carefully and to resist passing fads and advertisers' claims. It requires reuse and repair of old goods instead of discarding them for new ones. It also means refusing to buy items that put heavy burdens on natural resources.

Until recently, little attention was paid to the source reduction phase of waste management. A 1993 article in *National Wildlife* states, "The average person is likely more familiar with the idea of waste reduction than source reduction. In the semantics [vocabulary] of the world of trash, waste reduction has for years meant burning, recycling, or composting. Source reduction, on the other hand, means drying up the stream."

The throwaway culture

For most Americans, drying up the stream is a formidable challenge. The very nature of the American economic system sets up barriers to source reduction by emphasizing continuous growth and consumer spending. Advertising encourages people to buy the latest fashions in clothing, sporting equipment, automobiles, and thousands of other items. Widespread availability of credit and installment buying makes it possible for more people than ever before to own things.

This material abundance has led the United States to become the most wasteful nation on earth in terms of garbage generated per person. A major reason for this wastefulness is the American fondness for disposable

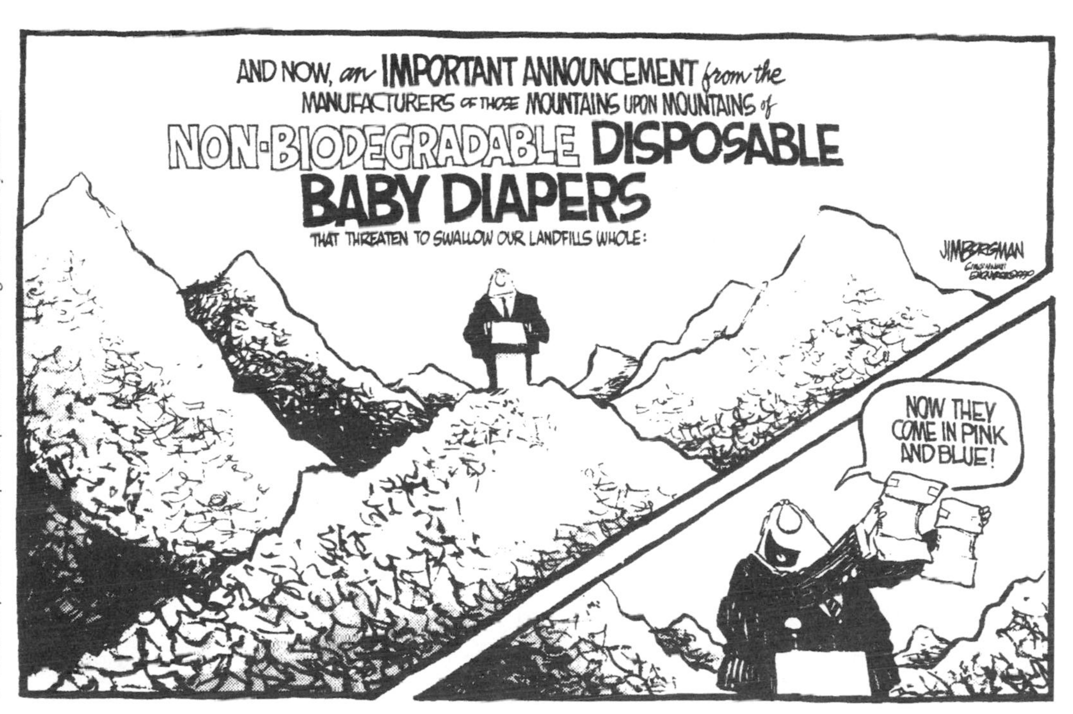

items. Valued for their ease and convenience, disposable goods constitute an important part of the American economy. Hundreds of limited-use items are purchased daily, including writing pens, razors, diapers, paper goods, and even contact lenses and cameras.

To most consumers, who never see the whole picture, it seems insignificant to toss a disposable diaper in the trash can or flip an old razor or pen into the wastebasket. But, as Jim Motavalli, editor of *E Magazine*, points out in a 1996 article, "We're chucking out 10 to 20 billion disposable diapers, two billion razors, 1.7 billion pens and 45 billion pounds of plastic every year."

America's throwaway attitude has profited those companies making disposable goods, but it has hurt other sectors of the economy. Many repair shops have seen a steady decline in business. Motavalli believes the reason for this is the wide availability of cheap, nonrepairable goods. In *E Magazine*, he tells of interviews with fix-it shop owners across the country. "Many of those interviewed say their work is increasingly embattled by cheap imported goods—which make fixing something broken more expensive than buying it new—and disposable designs that don't even allow some new products to be taken apart, let alone repaired."

One example is shoes. "Americans don't get their shoes fixed anymore," a shoe repairman says. "They just replace them. . . . They make these injection molded shoes in China . . . and because of the molding they can't be repaired. Even if they could be fixed, it would cost more than a new pair." Those who repair clocks, small engines, electric motors, radios, TVs, and clothing make similar assessments.

Packaging waste

An enormous amount of landfill waste today comes from the wrappings and containers in which products are packaged. In the old general stores that supplied people's needs in the early part of the century, many foods, such as beans, flour, sugar, crackers, and pickles, came in large and attorney, took a long leave of ab- ob to help develop the institute. "Dick and actice what they preach," Motavalli writes E Magazine article. "For 20 years, they ght disposables and recycle as much as possi- fill two trash cans a year.) They compost their ste, make do with one small car, and rarely buy thes." Like Dominguez and Robin, the Roys take ay from the institute, although they both work full- t it.

wnshifting is practiced not only by individuals; cer- types of businesses are also profiting from the trend ard source reduction. For example, many dairies and ier bottlers (even wine makers) are washing and refill- ing glass and plastic containers instead of purchasing new ones. In some communities, laundering cloth diapers has become a profitable business. Rental businesses are enjoy- ing an upsurge, as more Americans than ever before rent seldom-used tools and other items instead of buying them. Almost anything may be rented today from cement mix- ers to coffee urns.

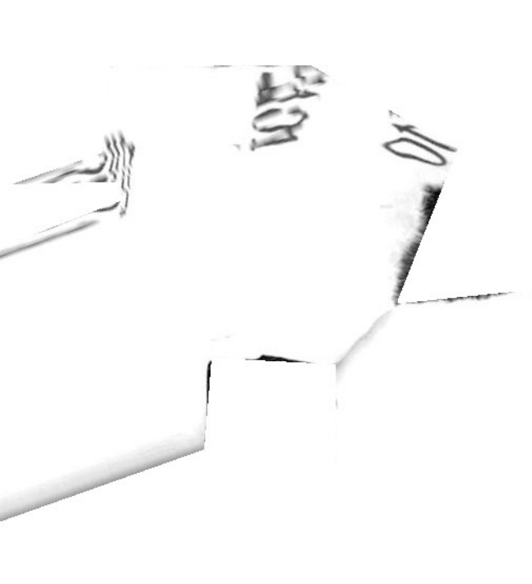

"Thrift shops are popular, too," a frequent customer says in a 1995 *E Magazine* article. "Buying inexpensive second hand items takes the responsible consumer out of the over-advertised, name-brand hype shopping loop. That doesn't mean you can't find designer goods in the thrift stores: Because the U.S. has so many wealthy and waste- ful people, goods with originally steep prices can be had for a song."

The significance of downshifting

Whether the voluntary simplicity movement will lead to significant changes in American consumption habits is a matter of disagreement among observers. A 1995 poll in *U.S. News & World Report* says, "48% of Americans say they have taken steps in the past five years that could sim- plify their lives." Other observers see this as only a tempo- rary phenomenon, or perhaps the result of something other than a desire to live simply.

In the first case, doubters say that
claim to be downshifting are young pe
haven't settled down yet. As they grow
children, their lifestyles will become more c
second case, those who say they are volun
shifting may actually be having financial diffic
tionalizing career choices that have not worked
"They say they are trying to simplify their lives," a
ogist says, "when in fact other forces have propelle
decisions."

Writing in *E Magazine*, Motavalli supports the
that simplified living is a significant trend, but, he
the battle has only begun:

> The fight for the hearts and souls of American consumers is just getting started. It took a towering figure like [consumer advocate] Ralph Nader to convince Americans that consumers had any rights at all, but now that the sleeping giant is awakened—and in a position to actually *withhold* its buying power—the world is beginning to take notice. If we can stem the seemingly inexorable [unstoppable] rush to constantly buy and consume more, the beleaguered planet will have gained some valuable breathing room.

Industrial ecology

In industry, source reduction ideas and practices are referred to as "industrial ecology." The term describes a new model for industry that has environmentalists excited and hopeful. "The end of the 20th century has seen a subtle change in the way many industries are confronting environmental concerns," writes Robert A. Frosch in a 1995 issue of *Scientific American*. "They are shifting away from the treatment or disposal of industrial waste and toward the elimination of its very creation."

This new model views industry as a whole rather than as isolated factories, refineries, mines, and mills. It is patterned after the natural ecosystem, which Frosch defines as "an integrated whole. . . . Nothing, or almost nothing, that is produced by one organism as waste is not for another organism a source of usable material and energy. . . . With this insight from the natural ecological system, we bulk containers. Clerks would fill customers' orders from barrels, boxes, and chests. Meat was sold in butcher shops and cut to order while customers waited. Fresh vegetables and fruits were grown locally or at home.

The growth of packaging evolved along with a new kind of store—the supermarket—in which customers help themselves to individually wrapped products from shelves, freezers, and refrigerated cases. According to the EPA, products should have only enough packaging to protect the product, preserve its freshness, or convey necessary information. Since packaging costs money, the question arises, Why do manufacturers use so much packaging material?

An overflowing wastebasket illustrates the wastefulness of modern American culture.

Some products are overpackaged for convenience, such as ease of transportation or stacking on store shelves. But most packaging is there simply to sell the product. Attractive packaging may add to the article's appeal or, sometimes, make a small product appear larger. Occasionally, both goals are accomplished at the same time. For example, a food processing company cited by an environmental watchdog group in 1995 sells lunches in attractive boxes that have much more box than lunch. Neither the box nor the wrappings are recyclable.

In a 1995 *Discover* article, science writer Jeffrey Kluger says consumer decisions not to buy such items (often called precycling) will quickly get the attention of manufacturers. "The answer for shoppers tired of buying as much plastic as product, as much beverage as box," Kluger writes, "is to begin putting their collective foot down and quit choosing the offending brands. The sooner manufacturers see that it is the less-dressed brands that make it off the shelves and the overdressed ones that get left behind, the sooner they'll start stripping them down to their packaging essentials."

The most publicized controversies over excess packaging involve restaurants, particularly those that serve fast foods. To dramatize this problem, Kluger itemizes the waste portion of his deli take-out cup of tea:

> Destined for the local landfill as a result of my decision to buy a single serving of orange pekoe were: one eight-ounce cup; one plastic lid; one used tea bag consisting of bag, six-inch length of string, two staples, and a festive Lipton tag; one six-inch plastic stirrer; three packets of sugar; two nondairy creamers; one wedge of lemon; three six-by-eight-inch napkins; one paper bag; one cash register receipt. Having always considered myself something of an environmentalist, I was stunned at this pile of beverage-related debris. What, I wondered, would the environmental impact have been if I had actually ordered food? Would a side of coleslaw have required Superfund cleanup legislation? Would a BLT have required the intervention of the EPA?

Fast-food servers don't have time to ask customers what they need with their tea or bagel, Kluger observes. They oversupply patrons to make sure all customers are adequately served. Many fast-food chains have responded to public criticism by changing their packaging and serving practices. Even so, the fast-food industry, which relies on disposable wrappings and containers, is an important aspect of American life and will no doubt continue to be a major source of waste.

Fast-food chains that use disposable wrappings for convenience have been accused of overpackaging, which critics say results in piles of unnecessary waste.

Downshifting

In spite of formidable obstacles, some headway toward source reduction is being made. On the consumer level, this is called downshifting. Downshifting means the voluntary decision of individuals and families to work less, spend less, and live more simply.

Motavalli observes in his 1996 *E Magazine* article that downshifting actually involves only a small portion of the population at present.

> For now, most Americans know they're dissatisfied with their lives, but find it difficult to "downshift" and change the consumption patterns that made them that way. It would take a powerful force to slow down America's consumerist engine, which has convinced a majority of American young people that it's "extremely important" to have at least two cars, the latest clothes, an expensive stereo and a vacation home.

There are some indications, however, that a downshifting trend may have begun. One indication is the popularity of several recent books on the subject, such as *Your Money or Your Life* by Joe Dominguez and Viki Robin, published in 1992. Both authors dropped out of fast-paced business careers to pursue other kinds of work that were less profitable, but more satisfying to them.

When other people became interested in their experiences, Dominguez and Robin began giving seminars, and eventually wrote their book. Not wishing to profit from their new careers, they established the New Road Map Foundation, to which the proceeds go. The goal of this Seattle-based foundation is to help people curb excess spending and live less complicated lives.

Although the New Road Map Foundation was not founded primarily to promote waste reduction, other organizations in this so-called voluntary simplicity movement do have source reduction as their guiding principle. One of these is Use Less Stuff, founded by Bob Lilienfeld in 1993 and described in a 1995 article in *U.S. News & World Report*:

> Two years ago, inspiration hit Lilienfeld when he helped a plastics company conduct a recycling campaign that was a

Old paint is collected to be renewed and donated to low-income families. Many people make efforts like this to reuse materials, as well as conserve resources and consume less overall.

> big success. He started to publish a newsletter and last month announced the first Use Less Stuff Day to encourage folks to conserve over the holidays. For instance, he says that in New York City, one less grocery bag per person per week would save five million pounds of waste per year and $250,000 in disposal costs.

Another organization advocating source reduction is the Northwest Earth Institute of Portland, Oregon, founded in 1993 by Dick and Jeanne Roy, a husband-and-wife team.

Buying cheap, easily replaceable items results in piles of garbage like this one. Alternatives to constantly consuming new merchandise include repairing products, renting items, or buying secondhand.

are beginning to think about whether there are ways to connect different industrial processes that produce waste, particularly hazardous waste."

As an example, Frosch describes an ecoindustrial park located in Kalundborg, Denmark, where several industries are located near one another in order to exchange waste products. Frosch says, "An oil refinery employs waste heat from a power plant and sells sulphur removed from petroleum to a chemical company. The refinery will also provide sulphur (as calcium sulfate) to a wallboard producer to replace the gypsum typically used. Excess steam from the power plant also heats water for aquaculture, while it warms greenhouses and residences."

While similar parks may prove unworkable in a large country like the United States, single industries also may benefit from industrial ecology. In a 1995 article in *Environment*, Frosch gives an example:

> A firm that polished metal parts by tumbling them with stones and water faced the problem of disposing of the water polluted by contact with the metal. It found that it could solve this problem by installing a settling and filtration system. Most of the water was reusable immediately; when the water remaining in the filtered sludge was removed by compression, the metal in the filter cake could be extracted by

> burning and resmelting. The process proved to be an overall money saver, and the capital investments required quickly paid for themselves.

The filtration process not only reused natural resources, it also eliminated the cost and dangers of disposing of hazardous wastes. One caution given in the article, however, is the cost-effectiveness of source reduction. Since energy is consumed in reprocessing as well as in processing, energy costs must also be figured into the equation.

Another aspect of industrial ecology focuses on developing manufacturing methods that are less destructive to the environment. A special advertising section by the National Association of Manufacturers in a 1995 issue of *Business Week* states:

> America's corporations have discovered the business advantages to adopting sound environmental practices that enhance productivity and product quality even as they vastly reduce or eliminate wastes that would otherwise go into our land, water, and air. 3M and Dow have the most established programs, but many others, including DuPont, G. D. Searle & Company, Eastman Chemical, and Cooper Tire and Rubber have also made environmental practices a cornerstone of overall corporate policy. . . . Companies have learned that controlling waste on the front end of the manufacturing cycle is a lot more efficient and ultimately less costly than controlling it at the back end.

Frosch reports that automobile companies are substituting water-based paints for those that contain large amounts of organic solvents, and the electronics industry is using substances in its manufacturing processes that are less toxic than those used previously.

Industrial ecology also holds manufacturers responsible for products throughout their life cycles. "Product stewardship offers clear benefits for the environment," Frosch says. "Preventing waste and pollution is a more natural impulse if firms retain a certain 'ownership' of their products—that is, if the responsibility for those products remains inside the system." Frosch notes that some industries are already beginning to design their products with life cycles in mind. Engineers now speak of design for recycling and design for environment as well as design for manufacturing.

Industrial ecology involves companies using one another's waste products in their own manufacturing processes in an effort to keep them from choking landfills or harming the environment.

As promising as these signs are, there are many wrinkles to be ironed out before industrial ecology can develop fully. For instance, new hazardous disposal regulations will be needed to allow one company's waste to be transported and reused by another industry. Greater openness

and cooperation among companies will also be necessary, a difficult step because many companies fear industrial secrets may be exposed through such exchanges.

Frosch admits industrial ecology hasn't gotten very far yet, but he is hopeful:

> The ideas and practices that fall under the heading of industrial ecology are still in their infancy. In all probability, only a small fraction of U.S. manufacturing companies are aware of them or have actually put them into practice. On the positive side, however, this group includes such major companies as Xerox, 3-M, Dow Chemical, AT&T, and the three largest automobile manufacturers in the United States, all of whom, in varying degrees, are attempting to apply the principles of industrial ecology to their businesses.

The problems remain

Regardless of how environmentally responsible industry may become, source reduction ultimately depends on consumers. Experience has shown that whenever there is a demand for a product or a service, someone will supply it. Whether Americans will modify their demands for material goods in order to reduce waste and preserve the earth's dwindling resources remains to be seen.

At present, more emphasis continues to be placed on disposing of garbage than on creating less of it. To deal with the continuing problem, scientists and researchers are looking at new ways to address the problems of waste management.

5

The Future of Waste Management

A STUDY OF the garbage problem in America makes it clear that there is no single cause. Neither is there a single solution that will work in all cases. Nevertheless, certain important themes emerge from today's public debate on waste management problems.

One of these is public realization that the days of "out of sight, out of mind" are over. Increasingly, waste is in sight, wherever we go, wherever we live. None of us can dismiss it or pretend it isn't our problem anymore. Another theme is the increased role source reduction and recycling must play in the future.

Up to now, municipal governments have generally preferred landfills and incinerators because they tend to be quick fixes. Investors and building contractors like them because they are profitable. Recycling has never been able to compete on the same level, but that trend may be changing. The EPA crackdown on landfills and public dislike of incinerators (not to mention their enormous cost) has caused many municipalities to revive the recycling option.

The authors of *Beyond 40 Percent* insist that recycling has the potential to become a major component in waste management, not just a sideline. Not only are high levels of recycling possible, but reduction and recycling must outrank other methods. "Preventing waste generation is recognized as the first priority in solving the waste crisis,"

These recycled shredded steel cans will be used as a fuel source. Reducing waste generation and recycling, as opposed to landfills and incineration, are now seen as viable solutions to the garbage problem.

the authors state. "Implementation of recycling and composting systems is recognized as the second priority. Incineration and landfill disposal are considered last resorts. . . . Materials recovery, not materials destruction, is the objective of grass-roots efforts toward solid waste management in a sustainable economy."

The same stand is taken by the Environmental Defense Fund as set forth in *Recycling and Incineration: Evaluating the Choices*. The first of seven steps for solving waste management problems states, "Waste reduction must be a top priority." The second is that city planners should give recycling a fair chance when deciding which waste management methods to use: "Any proposals to build new incinerators or landfills [must] include a full and fair comparative evaluation of the potential of waste reduction and recycling."

Nevertheless, even the most enthusiastic recyclers realize that reduction and recycling alone cannot cope with all of our waste problems, at least for the foreseeable future. Landfills, incinerators, and other remedies will still be needed for unrecyclable wastes and emergency cleanups of oil spills and hazardous waste pollution. The quest for new techniques to meet such emergencies and perhaps revolutionize present-day methods goes on in laboratories all over the world. Some new technologies in testing stages involving both biological and mechanical processes show promise.

Biological solutions to waste problems

Bioremediation is a biological technique for reducing waste. Today, it is being used on a variety of wastes from oil spills to sewage. Bioremediation uses bacteria to change waste materials into harmless substances. Bacteria are microorganisms usually associated with sickness and disease, but in bioremediation, bacteria act beneficially. A 1994 article in *American City & County* states, "On the list of things that people feel warm and fuzzy about, bacteria are somewhere between traffic jams and taxes. . . . But if they have never been thought of as man's best friend, bacteria are making inroads in that direction by providing safe, natural, cost-effective cleanups of water polluted with everything from fuels to heavy metals."

Polluted soil can also be cleaned up with bacteria, as was demonstrated at Prince William Sound, Alaska, after the disastrous *Exxon Valdez* oil spill in 1989. According to James D. Snyder in a 1993 article in *Smithsonian*, all sorts of methods were used to clean the oil from rock-strewn beaches, including jets of hot water and plain old hand scrubbing.

When none of these methods produced the hoped-for results, bioremediation technicians asked the EPA to let them try their techniques. A small test plot was set up on a rocky, oil-soaked beach and sprayed with a type of fertilizer commonly used on farms. Two weeks later, the test plot could be seen from the air as a much cleaner patch on the beach than an untreated control plot nearby.

Bioremediation advocates then sought permission to try special oil-eating bacteria, but, mindful of possible risks involved, the EPA moved cautiously. By the time the new tests were approved, the cleanup was almost over. "Yet the big Alaska spill was a watershed event for both the EPA and for bioremediation," Snyder reports. The EPA now approves bioremediation for certain kinds of pollution problems. Because it treats polluted areas where they are, bioremediation is much

Two workers prepare oil-eating microorganisms to clean up an oil spill. This technique of handling hazardous waste, whereby the bacteria convert the waste to harmless substances, is known as bioremediation.

cheaper than digging up polluted soil, hauling it away, and burning it in incinerators.

There are several companies in the United States that cultivate bacteria to attack or digest different kinds of waste. Snyder says, "One organism's poison is another's staff of life; the right microbes will eat just about any toxic waste." Microbes can also be made to work faster with soil aeration and the addition of nutrients.

Bioremediation is used on nonhazardous domestic garbage, too. Palm Beach County, Florida, "recently opened a $1.3 million facility to compost some 30 tons a day of sewage sludge and yard clippings," Snyder reports. "One of perhaps 20 like it nationally, the Palm Beach facility features two 250-foot-long trough-like bays under one roof. There, automation helps microbes munch out in minimum time. . . . The result, 21 days later, is an almost odorless, fluffy soil conditioner that the county uses on its own parklands and sells to garden shops."

Bioremediation in sealed-off sanitary landfills is not as effective as it is in open places because bacteria need air and moisture to work well. In order to utilize this important new technology, however, wetter landfills are being designed in some localities.

Like all garbage disposal solutions, bioremediation has its limitations. "Bioremediation is still in its infancy," Snyder concludes, "and there is no telling how far it can go. . . . Whatever happens, it appears a safe bet that microbes will be the Marines of future environmental cleanups."

Phytoremediation

Phytoremediation, another biological solution to waste problems, is a method of cleaning up contaminated soils by growing special plants on them. "Both amateur and professional botanists have long known that certain plants can concentrate metals to levels that would kill off most species," says a 1995 article in *Science*. "For example, for hundreds of years, prospectors looking for copper used this knowledge to hunt for precious ores lurking near the surface." This knowledge is now being used to remove many potentially hazardous substances from the ground at mines, military sites, chemical plants, and municipal dumps.

Although a large variety of plants are being tested for phytoremediation, even common domestic plants can be used. Corn and peas absorb lead when certain chemicals are added to the soil. Sunflowers are used to dispose of uranium waste at an Energy Department site in Ohio. "Cleaning it up would ordinarily mean carting away thousands of tons of polluted dirt at great expense," a 1996 *Business Week* article states. "Now, hope for a better, cheaper approach is blooming—literally. In a small greenhouse on the factory grounds, scores of sunflowers are sucking the uranium out of contaminated water pumped from the soil."

Plants that have taken up toxic elements are contaminated and must be dealt with accordingly. Sometimes they are harvested and sent to smelters where the metals are extracted. But even when the metals cannot be recovered, the contaminated plants take up much less space in hazardous landfills than polluted dirt would occupy.

Like bioremediation, phytoremediation is new and largely unproven, but a number of companies are refining the process today. Some are experimenting with genetic engineering, trying to produce plants that are "hyperaccumulaters," able to remove large amounts of various metals.

Vitrification

Other experimental waste reduction methods being tested today rely on chemical and mechanical processes rather than living organisms. One of these, vitrification, turns waste into glass by melting it along with batches of molten glass in extremely hot furnaces. A 1995 article in *Science News* reports, "Vitrification can work with almost any form of waste—from soil contaminated by lead or radon [a gaseous element produced by the decay of radium] to medical wastes, industrial sludges, and radioactive wastes."

One of the most promising uses for vitrification is turning nuclear waste into glass. While this will not completely solve the problem, it will go a long way toward safely containing radioactive wastes until more effective methods are developed. A 1995 article in *Barron's* describes a vitrification plant in South Carolina:

> In a huge clearing not far from one of the old factories [abandoned buildings where radioactive materials were refined in past decades] stands another giant concrete building. But here the concrete is fresh, and it offers hope, rather than fear. It's the Defense Waste Processing Facility. Operated by Westinghouse, this new plant will go into operation later this year [1995], processing highly radioactive material—spent fuel from military reactors and tanks full of radioactive chemical byproducts. It will concentrate the radioactive material, melt the concentrate together with borosilicate sand and pour the resulting glass into stainless steel containers. . . .
>
> This process, called vitrification, will not make nuclear waste safe and cuddly. The stuff will remain dangerously radioactive for at least 10,000 years. But the glass will be stable and easier to handle. If it's buried, it won't break up or leak radioactive material into the air or water.

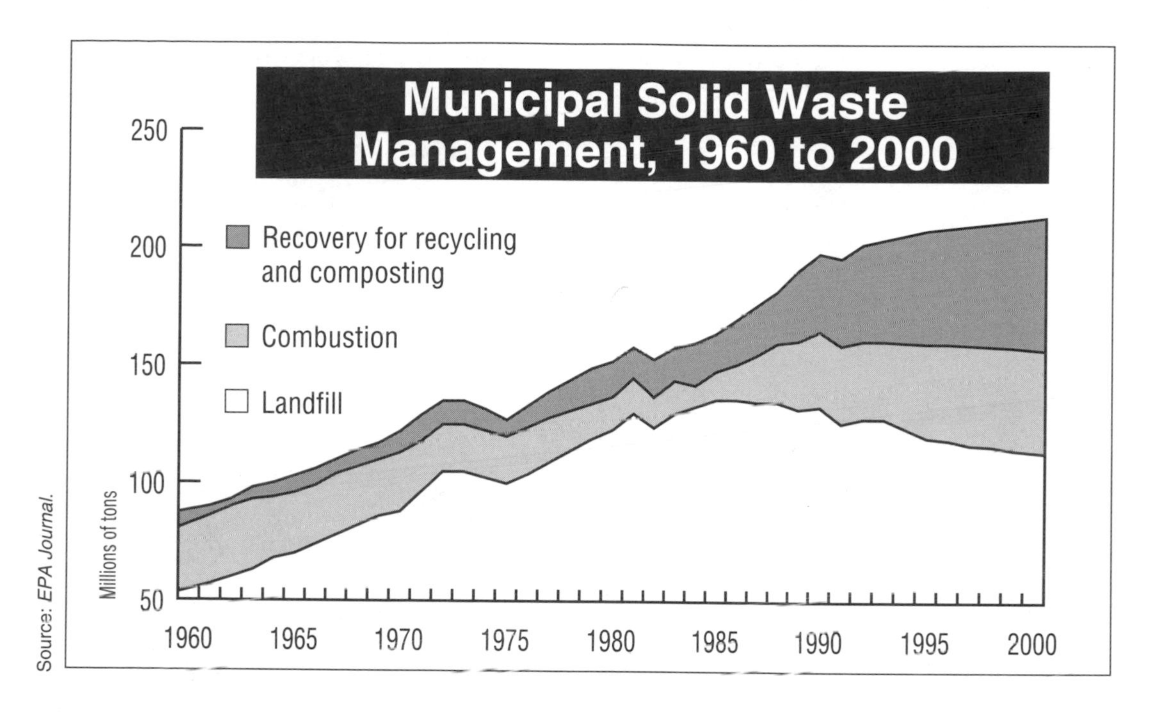

Source: *EPA Journal.*

Vitrification is also being used successfully with asbestos, a fireproof insulation material composed of fibers that can be very harmful to humans. Vitrifying asbestos makes it safe and greatly reduces the amount of space asbestos waste material takes up in landfills. Eventually, landfilling may be unnecessary. For instance, a company in Maryland plans to recycle vitrified asbestos. "Eventual uses will include fill for building and highway construction projects, ceramic bricks, and even insulation," according to a *Science News* article.

Vitrification has even given rise to artistic expression. Craftsmen who specialize in glass blowing are creating beautiful vases and other ornamental objects from vitrified asbestos.

A much hotter, more intense method of vitrification utilizes plasma or jet torches. The heat from plasma torches is so great, it is often compared to artificial lightning. The technology was developed in the 1960s for the space program when scientists had to find a way to duplicate the high temperatures space vehicles are subjected to when they reenter the earth's atmosphere.

Experimentation eventually led to the invention of the plasma torch, which can produce temperatures as hot as the sun's surface.

"[Scientists] pass a strong electric current through a rarefied gas, ionizing it to produce a flame that, at up to 8,000 degrees Celsius, is much hotter than any fire," explains a 1993 article in *Scientific American.* "Faced with such intense heat and denied oxygen, matter does not burn; it dissociates, in a process called pyrolysis. Toxic hydrocarbons break down into simple gases. Metals melt and disperse. Soil solidifies."

Plasma torch prototypes

Waste management engineers were quick to adopt plasma technology for vitrifying toxic waste or turning it into harmless, reusable substances. A 1991 article in *Natural Science* notes, "Many prototypes for the possible applications of plasma technology are in various steps of development around the world. The United States and Canada are developing prototypes that could utilize plasma technology to pyrolytically treat various types of waste, including municipal wastes, used automobile tires, used oil, hospital wastes, and nuclear wastes."

One of the leading advocates of plasma technology is Dr. Louis J. Cicero of the Georgia Institute of Technology. He thinks plasma torches will revolutionize domestic waste in the foreseeable future. "It's going to end landfills and incinerators," he says in a 1996 issue of the *Albuquerque Journal.* "It's going to turn municipal waste into a renewable energy source." A 1995 article in *Compressed Air Magazine* describes a plasma torch demonstration by Dr. Cicero:

> With a student's help, Dr. Lou Cicero . . . was ready to show a few visitors a new way to take out the garbage . . . or to put it another way, to turn garbage into gravel. Standing by a steel drum, 3 feet in diameter and 3 feet deep and filled with soil, he cautiously lifted what looked like a giant pistol and set the barrel end into a hole that had been dug in the center of the soil. Motioning the visitors to step back, he threw a switch.

> Instantly, ash and sparks erupted from the hole, accompanied by a high pitched roar. At the same time, the needle of a gauge shot off the scale, indicating the temperature had gone beyond 1000 degrees F. After a couple of minutes, Dr. Cicero climbed a short ladder and stared into the burbling hell he'd created. A flame hotter than the surface of the sun was melting the soil. Overall, it looked like a pool of molten lava.

Not only will plasma torches take care of future garbage woes, Cicero believes they will solve old landfill problems as well. Mounted on crawlers and lowered into landfills through bored holes, plasma torches will vitrify the garbage deep within the trash pits, Cicero claims. Since plasma torches reduce the volume of waste by 80 percent, landfill capacity will be greatly expanded.

Furthermore, the residue may be reusable. The only end product of plasma torch treatment, according to the article in *Compressed Air Magazine,* is "a glassy, inert [nonreactive], brick-like material that is reported to be harmless. Sometimes it's called slag or gravel or nuggets. The slag can be left in place on the landfill to seal the site, or more garbage can be piled on top. It can also be removed and used as gravel or brick in roadway projects."

The reality of plasma torch waste disposal as an alternative to landfills and incinerators may be closer than anticipated. City officials in Raton, New Mexico, are currently trying to raise $3 million to build a plasma torch waste facility for their municipality. If they succeed, it will be the first of its kind in the United States. "What we're looking for is a permanent solution to the solid waste problem that Raton is having," says a Raton city official in the *Albuquerque Journal.* "It's really difficult for a city the size of Raton [population eight thousand] to come up with continuing money for landfills."

As the citizens of Raton have learned, the major obstacle at present is cost. A September 1995 article in *Scientific American* says:

> The hurdle has been economic: plasma can vaporize nonhazardous waste for about $65 a ton, whereas landfilling costs less than half that amount. But as landfill space dwindles and

stricter environmental codes are adopted, plasma waste destruction is becoming more competitive. . . . It may be a while before toxic waste is a distant memory or before you can zap your kitchen trash into nothingness with the flick of a switch, but many researchers are betting that plasma waste destruction is becoming a reality.

Future solutions in space

Space age technology that created the plasma torch may someday help ease the waste burdens of earth in other ways. Everyone involved with waste disposal has probably dreamed from time to time of launching garbage into outer space. With all that room out there, it seems a perfect way to relieve the earth of its garbage burden. One obvious drawback is the expense involved, but even if this method became cheaper, it would still have problems.

There is already a tremendous amount of space junk orbiting the earth. An article in a 1994 issue of *Audubon* says, "Locked in orbit by the laws of physics, a ghostly armada of relic rocket ships perpetually circles the planet.

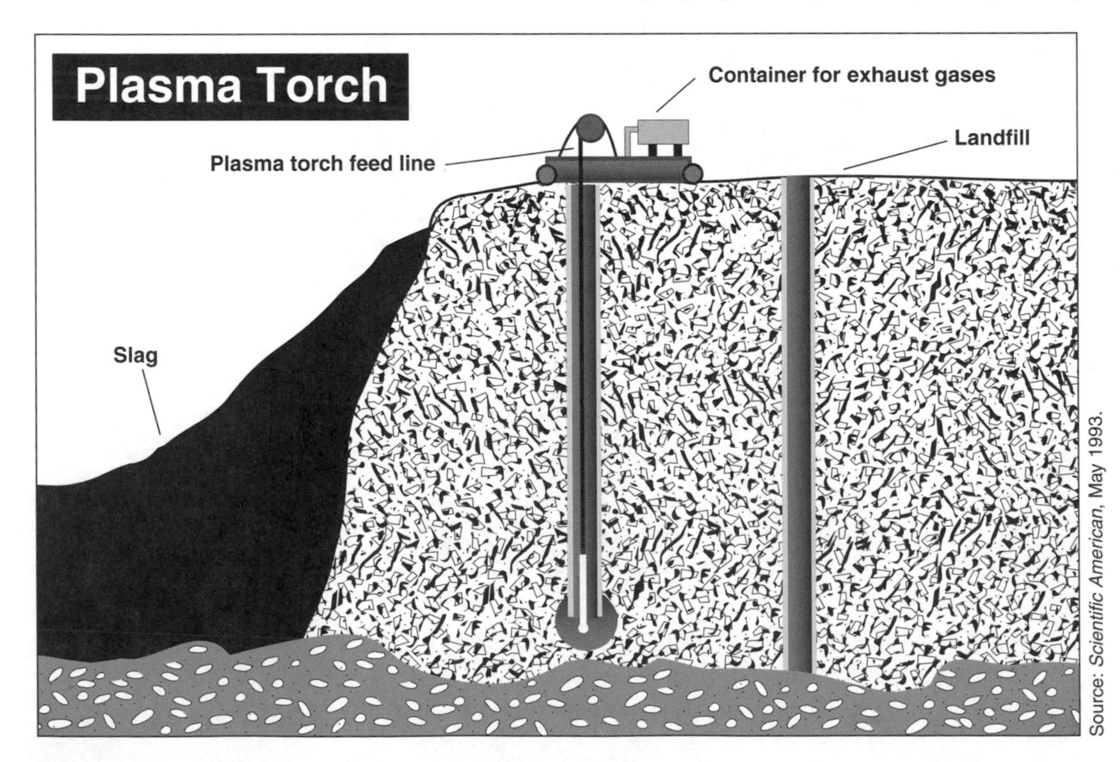

Nearly 3,000 tons of metal occupy this debris belt, most of it in the form of intact satellites; some contain small nuclear reactors. Even the space ships that travel to the outer solar system leave behind their booster rockets and other earth-orbiting junk."

Many of these objects have broken apart, creating thousands of small chunks of various sizes. Some of the larger pieces do not entirely burn up when they fall out of orbit, but plunge to earth in unpredictable places. In addition, there are billions of paint fragments and other tiny objects that have shaken loose from their parent machines. A tiny paint fleck hit the windshield of an orbiting space shuttle in 1983 and produced a small crater one-fifth of an inch wide. Fortunately, the shuttle's crew was unharmed, but the windshield had to be replaced at a cost of $50,000.

"To date, no accidental collision with orbital debris has claimed the life of any spacecraft crew member or destroyed any scientific instrument," the *Audubon* article reports, "but space agencies around the world claim it's just a matter of time before such a disaster occurs."

Of course, the purpose of sending garbage into space is not to keep it in orbit, but to burn it up. According to a 1995 article in *Scientific American*:

> The idea of loading toxic or other forms of waste on board a spacecraft and blasting them into the sun seems, at first glance, a nice solution to the earth's trash woes. At 5,500 degrees Celsius, the surface of Sol would leave little intact. But considering the amount of garbage each human produces—three to four pounds per day, on average—launches would simply be too expensive to conduct regularly. Add the possibility of a malfunction during liftoff, and space shots of waste seem impractical.

In *Space Garbage*, science writer Isaac Asimov suggests that advanced civilizations may get rid of their garbage in small black holes, which he calls the "ultimate trash-mashers." Theoretically, black holes are created by dying stars that are collapsing in on themselves. The mass of the dying star becomes so dense that nothing can escape its gravitational grasp, not even light.

Sending the earth's garbage into space to be burned up in the sun is one option for waste management, but the enormous expense it would entail makes it impractical.

Unfortunately, society has a long way to go before garbage can be banished into black holes. The idea of space as a future garbage disposal option cannot be dismissed, but it is not something on which today's solid waste managers can pin their hopes.

The realities of waste management today

None of these experimental methods, creative as they are, resolve the problem of protecting the earth's sources from overuse. On the contrary, they may magnify it. If quicker, cleaner ways to dispose of trash are invented, fewer people may feel the need to recycle, reuse, and reduce. Moreover, future solutions cannot solve present-day garbage problems. While the search for improved methods continues, waste managers must rely on the technology already in place.

Glossary

biodegradable: Materials that are capable of being broken down by microorganisms into simpler substances, which then may be used for other natural purposes.

bioremediation: Cleansing contaminated water and soils with bacteria. Bioremediation is a natural process, but it may be speeded up by using specialized bacteria and nutrients. Often used to treat oil spills.

composting: Turning organic material into fertilizer by allowing it to ferment in the presence of air and moisture. In the process, the original materials are broken down into simpler substances by earthworms, microbes, and insects.

downshifting: The voluntary act of reducing consumer spending and choosing a simpler lifestyle.

ecoindustrial park: A cluster of factories that exchange waste products beneficial to one another. It is based on a natural ecological model in which the waste products of certain organisms are life-sustaining substances for others.

EPA: The Environmental Protection Agency was authorized by Congress in 1970. The EPA administers nine comprehensive environmental protection acts such as clean air, clean water, RCRA, and Superfund.

garbage: A general term for household and business waste. Called municipal solid waste by the Environmental Protection Agency, it does not include industrial waste.

hazardous waste: Any kind of waste material in any form (solid, liquid, or gas) that threatens the well-being of the environment or the health of living things.

HDPE plastic: High density polyethylene; this is the type of plastic used to make such items as milk jugs and detergent bottles.

incinerator ash: The residue left after garbage is burned in incinerators. It is of two types, fly ash and bottom ash. The fly ash, collected from smokestack pollution control devices, is more toxic than bottom ash, which simply falls to the bottom after burning.

industrial ecology: A new model for industry patterned after nature in which nothing is wasted. The goals of industrial ecology are to recycle manufacturing wastes, use more environmentally friendly materials, and design products capable of being recycled.

leachate: Liquids that seep from rotting garbage.

mass burn system: A type of incinerator in which unsorted garbage is burned either to get rid of it or to produce energy for other purposes. Everything is burned with the exception of certain large items. Most incinerators in the United States are of this type.

methane gas: An odorless, colorless, flammable gas, one of the by-products of rotting garbage. Unvented, it can build up in landfills and cause explosions. It can also be captured and used to produce energy.

municipal solid waste: The EPA term for domestic garbage produced by households and businesses. It does not include industrial wastes.

PET plastic: Plastic made from polyethylene-terephthalate. PET is a clear plastic used to make soda bottles and other items. It may be recycled into fibers which are used to produce a great variety of products including clothing and upholstery.

phytoremediation: An experimental method for cleansing contaminated waters and soils with plants (phyta). Certain plants are able to take up harmful substances such as lead, cadmium, uranium, etc., without harm to themselves. The plants are then destroyed by burning, or smelted to extract the substances they have removed from the soil.

plasma torch: A device producing intense heat generated by passing a strong electric current through a rarefied gas.

Used on hazardous or domestic waste, the heat breaks down the waste into harmless gases and a much-reduced volume of reusable slag.

RCRA: Resource Conservation and Recovery Act, passed by Congress in 1976 to amend the Solid Waste Disposal Act of 1965. The RCRA distinguished hazardous from nonhazardous waste for the first time, and set forth different regulations for handling each category. It also recognized recycling and conservation of raw materials as goals of waste management in addition to disposal. Administered by the Environmental Protection Agency, it has been amended several times.

refuse derived fuel system: A type of incinerator in which recyclable or unburnable materials are first sorted out. The remaining garbage is shredded and compacted into fuel pellets which are used to produce energy at the plant or sold to other energy customers.

source reduction: Reducing the volume of garbage by not producing so much in the first place. Source reduction (also called resource reduction) may be practiced by both consumers and manufacturers. It is now considered top priority by the EPA and many other organizations involved in solving waste problems.

Superfund: The nickname for the Comprehensive Environmental Response, Compensation, and Liability Act (CERCLA). This act sets aside millions of dollars to clean up critical toxic sites without first having to determine who is at fault. Once the responsible parties are identified, however, the act allows the government to recover cleanup costs from them.

toxic waste: Hazardous waste that can cause serious illness or death.

vitrification: Turning substances into glass by means of intense heat and the introduction of certain chemicals.

waste-to-energy plants: Incinerators that use garbage to produce steam or electricity.

Organizations to Contact

There are literally hundreds of organizations today whose agendas include some aspect of waste management. The groups listed give waste management high priority and produce educational materials about waste disposal, recycling, and source reduction.

Government

Environmental Protection Agency (EPA)

This agency was established by Congress in 1970 under the executive branch of the federal government. Its purpose is to oversee and coordinate the work of all federal committees and offices having to do with solid and hazardous waste disposal and air and water pollution. The EPA is also responsible for carrying out pollution control legislation passed by Congress. Ten regional offices operate throughout the United States.

The EPA publishes free educational materials and other public documents pertaining to solid waste. Call or write for the *Catalog of Hazardous and Solid Waste Publications.*

For written requests:
RCRA Information Center
Office of Solid Waste (5305)
U.S. Environmental Protection Agency
401 M St. SW
Washington, DC 20460

For telephone requests:
Call the RCRA Hotline, (800) 424–9346, Monday through Friday, 8:30 A.M. to 7:30 P.M. EST. On the voice mail options, press 1 for the documents department.

Nonprofit Foundations

Nonprofit foundations are those that promote special programs and causes through grants, gifts, and membership fees. Many foundations also sell merchandise, but the profits must be used to support the group's work.

Environmental Hazards Management Institute (EHMI)
10 Newmarket Rd.
PO Box 932
Durham, NH 03824
(800) 446-5256
fax: (603) 868–1547
e-mail: ehmiorg@aol.com

EHMI is dedicated to solving hazardous and other kinds of waste problems through education and relationship building. It bridges the gap between industrial companies and environmental groups, which are sometimes at odds with one another. Among EHMI's educational tools are videos, posters, activity books, and "wheels," on which students may "dial up" information about composting, recycling, and other waste problems. A free catalog (with price list) is available.

Keep America Beautiful
1010 Washington Blvd.
Stamford, CT 06901
(203) 323–8987
Web site: http://www.kab.org

Originally founded to combat littering, this organization has expanded its scope and now publishes a curriculum guide and other materials on a variety of waste-related subjects; some are available at no charge. Call or write for a free catalog.

National Recycling Coalition
1727 King St., Suite 105
Alexandria, VA 22314–2720
(703) 683–9025
fax: (703) 683–9026

This organization is made up of individuals and companies who promote recycled goods. It has initiated two national recycling campaigns—"Buy Recycled" for consumers and

"Buy Recycled Business Alliance" for companies. Write or call for fact sheets about the coalition's recycling programs.

Trade Associations

Trade associations represent the interests of companies engaged in the same (or similar) types of production. Many of the following conduct waste management research and sponsor recycling activities.

Aluminum Association
900 19th St. NW, Suite 300
Washington, DC 20006
(202) 862–5100
Web site: http://www.aluminum.org
This association provides free fact sheets about recycling aluminum. It also has videos for loan. Contact the association for details.

Glass Packaging Institute
1627 K St. NW, Suite 800
Washington, DC 20006
(202) 887–4850
fax: (202) 785–5377
Web site: http://www.gpi.org
Educators who contact this institute will receive a free packet of teaching materials about glass manufacture and glass recycling. Information and activities for K–12 students, as well as an attractive full-color poster, are included in the packet.

Steel Recycling Institute
680 Andersen Dr.
Foster Plaza 10
Pittsburgh, PA 15220–2700
(800) 876–7274
fax: (412) 922–3213
A packet of steel recycling brochures from cans to automobiles is available free of charge. Also free are activity sheets, posters, and stickers. Certain other items (such as curriculum guides and videos) may be ordered for a fee. An order form describing the materials is included in the packet.

Suggestions for Further Reading

Isaac Asimov, *Space Garbage*. Milwaukee: Gareth Stevens, 1989.

Jean F. Blashfield and Wallace B. Black, *Oil Spills*. Chicago: Childrens Press, 1991.

Judith Condon, *Recycling Plastic*. New York: Franklin Watts, 1990.

Walter H. Corson, ed., *The Global Ecology Handbook: What You Can Do About the Environmental Crisis*. Boston: Beacon Press, 1990.

EarthWorks Group, *50 Simple Things Kids Can Do to Save the Earth*. Kansas City: Andrews and McMeel, 1990.

Mark D. Harris, *Embracing the Earth: Choices for Environmentally Sound Living*. Chicago: Noble Press, 1990.

Barbara James, *Waste and Recycling*. Austin, TX: Steck-Vaughn, 1990.

Richard Maurer, *Junk in Space*. New York: Simon & Schuster, 1989.

Christina G. Miller and Louise A. Berry, *Wastes*. New York: Franklin Watts, 1986.

Joy Palmer, *Recycling Glass*. New York: Franklin Watts, 1990.

———, *Recycling Metal*. New York: Franklin Watts, 1990.

William Rathje and Cullen Murphy, *Rubbish! The Archaeology of Garbage*. New York: HarperCollins, 1992.

Allen Stenstrup, *Hazardous Waste*. Chicago: Childrens Press, 1991.

Ruth Taylor, ed., *Recycling Paper*. New York: Franklin Watts, 1990.

Michael Williams, *Planet Management*. New York: Oxford University Press, 1993.

Works Consulted

Periodicals

Nancy L. Abbott, "Voila! From Refuse Container to Composting Bin," *Biocycle*, March 1996, p. 57.

Tom Arrandale, "The Changing World of Landfills," *Governing*, August 1993, p. 59.

Attilio Bisio and Sharon Boots, eds., "Incineration," in *Encyclopedia of Energy Technology and the Environment*, vol. 3. New York: Wiley, 1995, p. 1,796.

———, "Waste to Energy Technologies," in *Encyclopedia of Energy Technology and the Environment*, vol. 4. New York: Wiley, 1995, p. 2343.

Christopher Boerner and Kenneth Chilton, "The Folly of Demand-Side Recycling," *Environment*, January/February 1994, p. 7.

Adrienne C. Brooks, "A Glass Melange: New Options for Hazardous Wastes?" *Science News*, January 21, 1995, p. 40.

Salvador L. Camacho, "Harnessing Artificial Lightning," *Natural Science*, December 1991, p. 310.

John Carey, "Can Flowers Cleanse the Earth?" *Business Week*, February 19, 1996, p. 54.

Compressed Air Magazine, "Barbecuing Matter," January/February 1995, p. 11.

Thomas G. Donlan, "Glass Houses," *Barron's*, June 12, 1995, p. 59.

Janet Essman, "Two-for-One Landfill Lease," *Conservationist*, February 1993, p. 46.

Robert A. Frosch, "The Industrial Ecology of the 21st Century," *Scientific American*, September 1995, p. 178.

———, "Industrial Ecology: Adapting Technology for a Sustainable World," *Environment*, December 1995, p. 16.

W. Wayt Gibbs, "Garbage in, Gravel out: Plasma Torches Transmute Waste into Harmless Slag," *Scientific American*, May 1993, p. 130.

David J. Hanson, "Hazardous Waste Incineration Presents Legal, Technical Challenges," *Chemical & Engineering News*, March 29, 1993, p. 7.

Dwight Holing, "Sounding the Smoke Alarm," *New Choices*, July/August 1993, p. 68.

Jeffrey Kluger, "Paper Trail," *Discover*, November 1995, p. 54.

Mike Kukuk, "Bioremediation: Bacterial Surf 'n' Turf," *American City & County*, October 1994, p. 40.

Laura M. Litvan, "Going Green in the '90s," *Nation's Business*, February 1995, p. 30.

John Marks, "Time Out," *U.S. News & World Report*, December 11, 1995, p. 85.

Bill McKibben, "Score One for the Mountain," *Audubon*, January/February 1994, p. 110.

Chaz Miller, "The Shape of Things to Come," *Waste Age*, September 1995, p. 60.

Anne Simon Moffat, "Plants Proving Their Worth in Toxic Metal Cleanup," *Science*, July 21, 1995, p. 302.

Jim Motavalli, "Enough!" *E Magazine*, March/April 1996, p. 28.

———, "The Real Conservatives: Whatever Became of Fixing Things?" *E Magazine*, July/August 1995, p. 28.

Roy Popkin, "Good News for Waste Watchers: Recycling, Composting Show Results for the Future," *EPA Journal*, Winter 1995, p. 18.

Janet Raloff, "Computer Redux: New Lives for Castoffs Is a Growth Industry," *Science News*, December 23 and 30, 1995, p. 424.

Scientific American, "The Ultimate Incinerators," September 1995, p. 180.

Seth Shulman, "Houses to Save the Earth," *Parade*, March 3, 1996, p. 4.

James D. Snyder, "Off-the-Shelf Bugs Hungrily Gobble Our Nastiest Pollutants," *Smithsonian*, April 1993, p. 67.

Dava Sobel, "Littering Space," *Audubon*, March/April 1994, p. 94.

Nadia Steinzor, "Striving for Justice," *ZPG Reporter*, March/April 1996.

Michael Terrazas, "Of Pencils & Procurement: Cities Closing the Loop," *American City & County*, January 1995, p. 44.

John Tierney, "Recycling Is Garbage," *New York Times Magazine*, June 30, 1996, p. 24.

Thomas M. Tierney, Director of Seattle Office of Management and Planning, private correspondence to author, June 17, 1996.

Dan Trevas, "Dump Your Own," *E Magazine*, May/June 1992, p. 9.

William Young and Bill Tikkala, "A Dump No More," *American Forests*, Autumn 1995, p. 58.

Books

Louis Blumberg and Robert Gottlieb, *War on Waste: Can America Win Its Battle with Garbage?* Washington, DC: Island Press, 1989.

Richard A. Denison and John Ruston, eds., *Recycling and Incineration: Evaluating the Choices*. Washington, DC: Island Press, 1990.

Newsday, *Rush to Burn: Solving America's Garbage Crisis?* Washington, DC: Island Press, 1989.

Susan Williams, *Trash to Cash: New Business Opportunities in the Post-Consumer Waste Stream*. Washington, DC: Investor Responsibility Research Center, Inc., 1991.

Index

About the Author

Eleanor J. Hall is a freelance writer who has had several careers. For many years she taught sociology and anthropology at a community college in southern Illinois. It was during this time that she first became interested in environmental causes. After retiring from teaching, she was for a time an educational counselor in an Illinois prison. From 1988 to 1993, she worked as an interpreter of history, museum specialist, and writer for the National Park Service at the Gateway Arch in St. Louis.

While associated with the National Park Service, she wrote two curriculum guides that were distributed parkwide: *The Oregon Trail: Yesterday and Today* and *When Two Worlds Met: An Observance of the Columbus Quincentennial*. For her curriculum guide on the Oregon Trail, she received the Midwest region's Tilden Award for outstanding performance in interpretation.

Her current freelance work includes *The Lewis and Clark Expedition* for Lucent Books, monthly children's activity columns in RV publications, and feature articles for various periodicals. She now lives, travels, and writes in the western United States.

Picture Credits

Cover photo: © Telegraph Colour Library/FPG International
AP/Wide World Photos, 16, 17, 19, 44, 51, 69
Archive Photos, 14, 48, 61
Archive Photos/Swoger, 13
© 1989 Neal Cassidy/Impact Visuals, 28, 30, 31
Corbis-Bettmann, 39
David Greenstein/Corbis-Bettmann, 67
© Carolina Kroon/Impact Visuals, 42
© 1990 Andrew Lichtenstein/Impact Visuals, 15
© 1991 Piet van Lier/Impact Visuals, 37
NASA, 82
© Cindy Reiman/Impact Visuals, 62
© 1992 Jeffry Scott/Impact Visuals, 64
UPI/Corbis-Bettmann, 23, 46, 54, 72, 73
© Jim West/Impact Visuals, 26, 49, 57